Jewels of Time
The World of Women's Watches

Jewels of Time

The World of Women's Watches

by Roberta Naas

THE CURATED COLLECTION

TABLE OF CONTENTS

- 6   FOREWORD

- 10   CHAPTER ONE
  Turn of the Century

- 46   CHAPTER TWO
  Artistic Eloquence
  - 52   Enchanting Enamels
  - 62   Sculptured Beauties
  - 68   Mystical Dragons
  - 76   Birds in Flight

- 100   CHAPTER THREE
  Architectural Influences
  - 102   Time Around the World
  - 108   Art Nouveau & Art Deco
  - 130   Mechanics of Time

- 168   CHAPTER FOUR
  Design Inspirations
  - 170   Icons in Fashion
  - 186   Gems of the Earth
  - 204   Royal Hues: Purple
  - 214   Dazzling Diamonds
  - 228   Shapely Sensations
  - 238   Floral Masterpieces
  - 256   Call of the Wild
  - 288   Secrets in Time

Chopard created this Chopardissimo precious watch in 1997. It features three heart-shaped diamonds: a 15-carat pink one, a 12-carat blue one and an 11-carat white diamond. They rest on a bracelet entirely set with yellow and white diamonds designed like flowers. The entire piece features 874 diamonds, weighing 163 carats and in 1997 was worth 25 million dollars. The three hearts open and close to reveal a "secret" watch dial entirely paved with yellow diamonds. More than 2,000 hours of work went into the making of this single watch, perhaps the world's most expensive jeweled piece.

FOREWORD — Women, Watches & Wonders

My experiences for the past quarter of a century plus as a journalist in the watch world began as a woman pioneering in a predominantly man's world. When I started, there were no other female watch journalists, there were hardly any female watchmakers or executives of brands, and there was resistance to the concept. Luckily, I, like many women in history in their own areas of expertise (large or small), was persistent. I was met with many a closed door and "old boys club" attitude before I earned my stripes through responsible reporting, and finally got to go behind the scenes of the most hallowed halls of watchmaking and interview the world's finest creators. Today, I am thrilled to say that there are other women watch journalists, and watchmakers, artisans, and executives of top brands — many of whom have commented on the pages herein about time, women, and women's watches. Today, women of the world increasingly spend their days, hours, minutes and seconds multi-tasking, often torn between being power-house career women, sensual wives and lovers, and nurturing mothers. // Just as I pioneered in a then-man's world, the 20th century was one where many women pioneered — it was a century of exploration, of growth in literature, arts and culture. It was a century of civil rights, gay rights and women's rights. This was a time when women finally got to hold center stage and simultaneously when wristwatches were enjoying their birth and growth. The parallel growth of women's roles and wristwatches throughout the 20th century is an interesting and intense one — one in which, as the century progressed, women's watches played an ever more vital and visible role.

This Piaget Limelight Party. Disco Ball inspiration watch is crafted in 18 karat white gold and set with 123 brilliant cut diamonds on a black satin strap.

Indeed, many of the world's finest watchmakers have spent a century (or more) creating timepieces dedicated to the supposed fairer sex. These watch brands understand that women want something all their own: a powerhouse timepiece that is also a delicate beauty — a work of art on the outside that has a roaring engine inside. These watch executives, designers, artisans — many of them women themselves — understand the complexity of a woman's nature and celebrate that in their creations.

Just as a woman is multi-talented and multi-sided, so, too, are women's watches. They are all about art, craftsmanship, beauty and precision. They are poetry in time and life. Few people realize that to unveil a new mechanical watch takes years of planning, research and development. The actual building of the movement inside the small case, usually only one or two inches in diameter, is another amazing feat. Often that movement is composed of hundreds of tiny hand-engraved, hand-assembled pieces. The assembling, alone, can take weeks for one watchmaker to complete. The end result is a harmonious wonder that, sadly, no one ever views, unless the watch has a transparent caseback or a dial aperture. But it works, nonetheless — ticking tirelessly daily, performing its many tasks unerringly. A woman's watch knows its master and dances to her tune as if like magic, thanks to the collective sciences and arts that go into its making.

In addition to the awe-inspiring, demanding work of the mechanical movement, many women's watches feature amazing dials that are artistic canvases, or gem-set cases, or bezels that are irresistibly seductive. These pieces can take months to create and are well worth the wait, as they reflect some important facet of art, style, and taste. Often influenced by life around us in their design, these timepieces are jewels as influential, sensual and powerful as the women for whom they are built. The innovation in their design is limitless, stopped only by one's imagination.

> "The greatest gift God has given me, other than my children, is time. Time to do what matters, to do the right thing, to live the right life. I time it all with my myriad of wonderful watches, and I don't like to waste any of it."
> — Roberta Naas

I have visited many of the finest watchmaking factories in the world, and I never tire of witnessing the intricate work that takes place therein. A centuries-old craft transpires in quiet solitude, side-by-side with cutting-edge technology. It is a profound sight and often one wonders what makes these artisans and watchmakers tick; what makes them so passionate about their craft? *Jewels of Time* is a book about their works of art and craftsmanship, unusual and beautiful statements of timeless history inspired by and created for women.

Roberta Naas in highlights of a lifetime of her watch expeditions around the world.

CHAPTER ONE
TURN OF THE CENTURY

It was in the late 1800's that women began taking more active, public roles in society — daring to walk more amongst the men. Not only did they take part in outdoor sports previously reserved for men, such as hunting and equestrian riding, but also they became more verbose on public platforms. As the 19th century turned into the 20th century, and the suffragette movement took hold in Europe and America, women evolved into outspoken, brazen beauties of pioneering spirit and multi-faceted dimensions. // In the early 1900's, socialites and royalty made headlines with their independent ways, philanthropic ventures and strong support of other women. More and more, individual women pursued their dreams. By the early 1920's, women were going where none had gone before: swimming the frigid waters of the English Channel, joining the Olympics, making statements about politics and presidents.

à la Montre Omega

"Do not squander time for that is the stuff life is made of."

— Benjamin Franklin

"Just like their male counterparts, women also want watches that are heirlooms for the future generations. It's maternal instinct – plain and simple. Unlike men, there are many more choices – from necklaces, earrings, rings to haute-couture gowns and to-die-for handbags. That is why it is all the more important to design women's watches with full consideration of aesthetics and a functionality that evokes their sense of value. Apart from stunning beauty or innate elegance in the watch, good functions that do not harm aesthetics (or nails, or eyesight!) would go down well for ladies with an impeccable sense of taste and choice."

— Chai Schnyder, Chairwoman, Ulysse Nardin

"Women, my friend, women! They want to control their own time too, and not with these ridiculous men's turnip watches! Thanks to miniaturized movements and the popularization of the wrist version, watches will become real bracelets, true pieces of jewelry. Look at how women are asserting themselves now!"

— Paul Mercier shared these thoughts with William Baume in 1919

Baume & Mercier diamond-set white gold Galaxy watch, circa 1973

A time when women finally got to hold center stage...

(Above:) Corum Chinese Hat, circa 1958
(Right:) Corum Chapeau Chinois, circa 2010

At the same time, the wristwatch industry had been growing up. While the first women's Swiss watch had actually been completed in 1827 for Queen Marie Antoinette by Breguet (though it was completed 34 years after she was executed so she never saw it), and a few other individual occurrences had been logged by specific brands of creating a "watch on a bracelet" for certain clients in the mid- to late-1800s, wristwatches had their official start in the 20th century. The first wristwatch in serial production was created by Cartier and had made its appearance in 1904. Since then, wristwatches fast became the most fashionable and functional statement.

"Women's liberation is the liberation of the feminine in the man and the masculine in the woman." — Corita Kent

(Above:) Mercedes Gleitz was one of the first women to partake in bold men's sports in the 1900's. In 1927, she swam the English Channel wearing a Rolex watch. When she emerged from the frigid waters, the watch functioned perfectly. It was named the Oyster for its water resistancy. (© AP Images)

(Right:) The Rolex watch Gleitz wore on her English Channel swim was an Oyster Perpetual. After the swim, the watch was inscribed on the back with her name and commemoration for the event. "Vindication Channel Swim, October 21st 1927."

These beautiful time-keeping wrist instruments made extraordinary progress in craftsmanship and technology during the 1900's, demonstrating many firsts that are exactly aligned with the advancements of women of the 20th century. In fact, when Mercedes Gleitz swam the English Channel, making headlines in 1927, she wore a Rolex water-resistant watch. The watch was found to have kept perfect time when she emerged 10 hours later from the icy water, and was then dubbed the "Oyster." A decade later, when Amelia Earhart made her daring solo flight as the first woman to cross the North Atlantic, she wore an Omega wristwatch. Not only did wristwatches play a role in women's exploration, they were important in society.

Aviator Amelia Earhart was the first woman in the world to cross the North Atlantic by air, solo, in 1932. She wore an Omega wrist chronograph tachymeter that featured an enamel dial and blued steel hands with a mono pusher chronograph crown at 2:00. In 1936, as she planned her around the world flight, she equipped her plane with an Omega dashboard chronograph and donned Omega watches for herself and her navigator. The aircraft and crew went down in the Pacific Ocean in July 1937.

"Nothing happens unless first a dream."

— Carl Sandburg

Queen Elizabeth II wore a Jaeger-LeCoultre Reverso 101 watch — the smallest mechanical wristwatch in the world — to her coronation in 1953.

Chopard first unveiled its now iconic Happy Diamonds collection in 1976 and won the Golden Rose in Baden Baden for the revolutionary new diamond setting design with free floating diamonds between the dial and the crystal.

Up until this point, wristwatches had been mechanical, with hundreds of parts in their movements. Certainly year after year, decade after decade, watchmakers strove to create thinner, more reliable watches — moving from mechanical watches to automatic winding timepieces (that wind via an oscillating weight inside that is powered by the wearer's movements). In the early 1970's, however, the quartz watch era went from dawn to full day — nearly decimating the unprepared Swiss watch industry. Many brands slowed their production of women's watches, believing women no longer wanted mechanical timepieces and not yet being able to deliver quartz. (While the Swiss had mastered the quartz technology, they had abandoned the effort to bring it to the market first. It was the Japanese who first introduced an analog quartz watch line on Christmas Day 1969.) Later, with the advent of the fashion-forward Swatch watch in 1983, Switzerland embraced quartz technology, and many of the luxury brands finally followed suit — believing women wanted the smaller watch size that was achievable with a quartz movement.

This Piaget watch with jade dial and Guilloche Milanese gold mesh bracelet was once owned by Jackie Kennedy. Aside from its fashionable design, it houses an ultra-thin Piaget 9P mechanical movement, circa 1965.
(© Publi Conseil / Piaget)

Beginning in the mid-1960's, the House of Piaget once again demonstrated its acute awareness of the world and trends around it — and delved fully into the realm of color. The house turned to the Earth's most wonderful stones to create dials and accents on watches that were innovative and astonishing — and definitely captured the mood of the era. Coral, lapis lazuli, turquoise, opal, tiger's eye and more became the gems of choice — and Piaget lit the way for a new age of collectible, individualistic watches. Piaget's artistic creativity shines in this 1971 example of a bracelet-watch from the "Slave Collection" with an oval dial of coral, enhanced by coral medallions.

(Above:) Bulgari "Snake" bracelet watch with head in a pavé of brilliant-cut diamonds, cabochon ruby-set eyes and a smooth-scaled body in gold, circa 1967

(Right:) Elizabeth Taylor wore this Bulgari Serpenti watch in the film Cleopatra in 1968. (© AP Images)

"The value of life lies not

In the 1940's Bulgari revived the serpent motif, using it for a watch for the very first time. First examples were bracelet watches highly styled with coils produced in the Tubogas technique or in gold mesh that wrapped around the wrist. These became iconic designs that are still marveled at and created today. The mini Bulgari Tubogas was created in 1960; the Serpenti rendition is from the same era.

"... in the length of days but in the use we make of them."

— Michel Eyquem de Montaigne, French Essayist

"Every detail has a purpose, a function. This technical exactness is expressed through pure lines and timeless elegance." — Bruno Belamich, Bell & Ross

The 1980's finally witnessed a return of fine Swiss watches for women, with some new brands emerging and exciting new materials making a statement on the market. Fine watches made their debut clad with rubber straps or high-tech ceramic crafting. Designer watch lines came into their own rights, and watches worn as fashion accessories were all the rage. Indeed, the last two decades of the 20th century were all about pretty watches. Colorful watches emerged strong, dials took on new dimension, and diamonds as adornments took center stage. The women's watch arena blossomed.

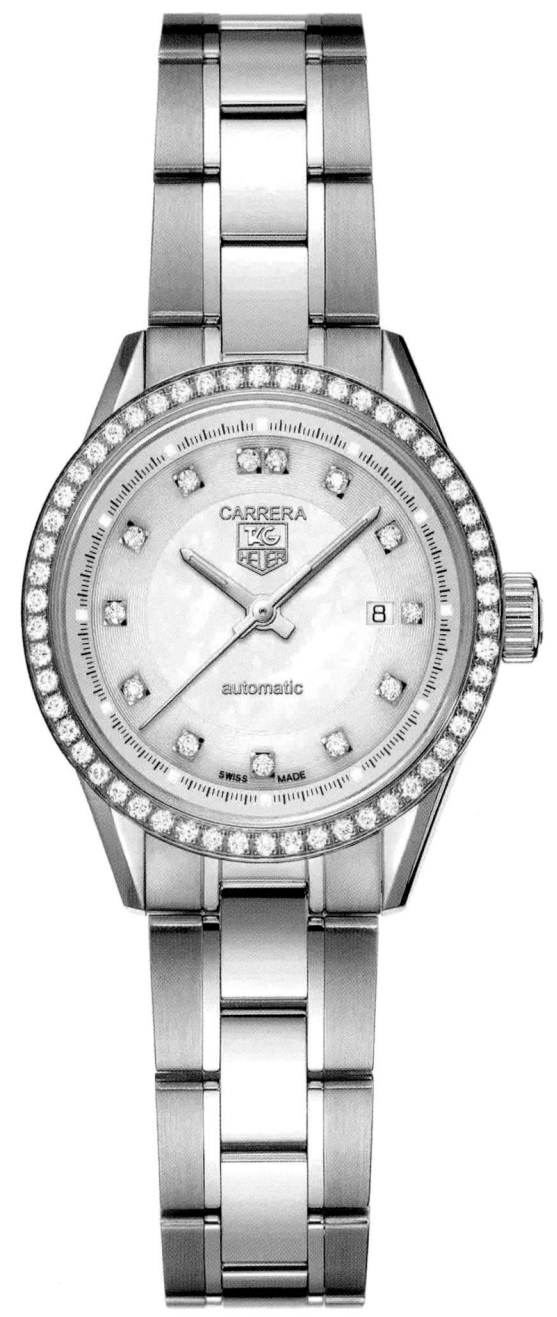

(Left:) BRS White Ceramic Phantom with Diamonds
(Right:) TAG Heuer Carrera Automatic

(Top left:) Women watchmakers make components and assemble Ulysse Nardin timepieces in the La Chaux-de-Fonds factory in Switzerland. © Ulysse Nardin (Top right:) Shu Qi, the famous Taiwanese-born movie star and glamorous personality, designed the new Amour Ladies Automatic Collection in close collaboration with Frederique Constant watchmakers in their Geneva Manufacture. © Frederique Constant (Middle left and right:) Master craftsmen build watches by hand in Cartier's own manufacturing operation in Switzerland. © Olivier Ziegler / Cartier (Bottom left:) Quality control is one of the key processes in Corum's La Chaux-de-Fonds factory. © Corum/Eveline Perroud (Bottom right:) Since Corum's founding in 1955 in La Chaux-de-Fonds, Switzerland, movements have been assembled by the brand's watchmakers, such as this T-bridge here at a state-of-the-art pressurized bench. © Corum / Eveline Perroud

Women lead many of the design and technological teams for some of horology's top brands today — a refreshing change from a century ago. These women not only design watches, but also design movements and are intimately involved in the research and development of technologically advanced nuances in time, many of which will make their mark in history. In the watchmaking halls, from building a movement to finishing and polishing, women have a dexterity and patience that are key. Additionally, some of the finest artisans in the trade — from engravers to enamellers — are women who bring with them their artistic eye and keen sense of beauty. Women in watchmaking lend a new view to time.

SEAH™ watches were created by a woman and were seven years in the making. The astrological timepieces stem from a desire by the creator and designer to reconnect people with their inner personality traits, strengths and oneness with nature — a sign of the times of the early 21st century when people are eager to return to their roots.

"I think women see time as a creation. Everything they do in life is a creation. From having a baby, which is creating life, to nurturing children, relationships, emotions, their time is a creation. For a woman, a watch is a statement of her connection to the world."

— Rachel Levy, Founder, Designer, SEAH(R) Watches

"We know women are the essence of time and without them, time would not flow as beautifully as it does."

— Sabine Rochat, Deputy Design Director, Baume & Mercier

Baume & Mercier's Linea, with domed crystal and rounded case, emulates femininity.

The watch features an interchangeable bracelet system and is set with 42 diamonds on the bezel and 36 diamonds on the first attachment link for a total of 1.4 carats.

Audemars Piguet Millenary Starlit Sky self-winding watch with center seconds, moon phases, power reserve indication and date in an 18-karat white gold case entirely set with 392 diamonds.

As the 20th century gave way to the 21st century, watchmakers once again took a closer look at their female clientele — reassessing lifestyles and desires. Today, brands offer timepieces that are as diverse as the women who are wearing them. Complex women, with their powerful jobs and busy lifestyles, seem to embrace the concept of the mechanical watch as well as the quartz watch. Today's watchmakers indulge. They offer complexities for women that men have long enjoyed in their watches — functions such as perpetual calendars, dual time zones, chronographs. They offer magnificent designs, rare and exceptional works of art, and the unexpected at every turn. In fact, they offer women Jewels of Time.

David Yurman Cable Collection Timepieces are crafted in 7mm sterling silver cable bracelet with 18-karat gold accents. They are set with an array of gemstones that include blue topaz, pink topaz or citrines. Dials are mother of pearl with diamond markers.

"Timepiece design marries creativity with precision industrial processes. Like any relationship, it requires a delicate balance and infinite patience."

— David Yurman

CHAPTER TWO — ARTISTIC ELOQUENCE

Wristwatches are so much more than timekeepers. They are horological works of technology and elegance, design commitments and personality statements that symbolize power and progress. For the finest watch brands in the world, creating a watch isn't just about putting numbers on a dial — it is about taking amazing creative license. For those innovative leaders, each watch is a canvas — an expression of singular beauty. While watchmaking itself is an art — one handed down from generation to generation for centuries — certain brands have taken the concept to elevated heights. Indeed, in the world of haute horlogerie, often such a harmonious blend of craftsmanship, history and culture exists that the result is a timeless masterpiece akin to the most coveted and revered museum painting. // Throughout this past century many brands have trained artisans to transform art from classical mediums such as enamel, cloisonné, mosaic, marquetry and even sculpted gold — onto tiny watch dials — offering breathtaking, unparalleled works of art — and catapulting the woman's wristwatch into a true statement of eloquence and elegance.

Artistically expressed both inside and out, this Cartier Crocodile Tourbillon houses the in-house caliber 9458 MC with flying tourbillon. The 167-part movement of this watch is inversed to allow for viewing, and the bridges are skeletonized for ultimate appeal. The diamond-set crocodile curls around the dial, protectively hovering over the coveted tourbillon escapement. (© Ines Dieleman / Cartier)

Artist and industrial designer Nathan George Horwitt is easily one of America's foremost designers of the 20th century. Born in Minsk, he moved, as a child, to New York and grew up in the Bronx. Influenced by the Bauhaus movement, Horwitt believed in sparse design. He created a frameless picture frame that was revolutionary for its time, and became a designer of furniture, china and interior spaces. In 1947, he created his own interpretation of time on a watch dial, based on the sun as it rotates Earth. The concept was defined by a single gold dot at "high noon" — the symbol that has become the iconic Movado Museum Watch. The round watch dial, devoid of anything but that dot and simple stick hands fulfilled his goal — to free the watch dial from unnecessary decoration. He received a patent for his design in 1958 and had three prototypes made. One was accepted into the permanent collection of the New York Museum of Modern Art in 1960 because of its powerful design. In 1960, Horwitt reached a partnership with Movado, which started creating wristwatches with Horwitt's dial aesthetic in 1960 — christening it the Museum Watch. Movado continues to create new and alluring evolutions regularly and to this day, the Movado Museum Watch remains one of the most legendary watches of the brand and the watch world.

"Without use there is not beauty — without beauty — what is the use?" —Nathan George Horwitt

This Movado Concerto® watch with the famed Museum® dial with signature dot at 12:00 is graced with subtle contours yet true to its original design aesthetic.

The first watch, unveiled in 1988, was the amazing "Andy Warhol Times 5" — a bracelet of five individual cases, each featuring a Warhol-photographed scene of New York City.

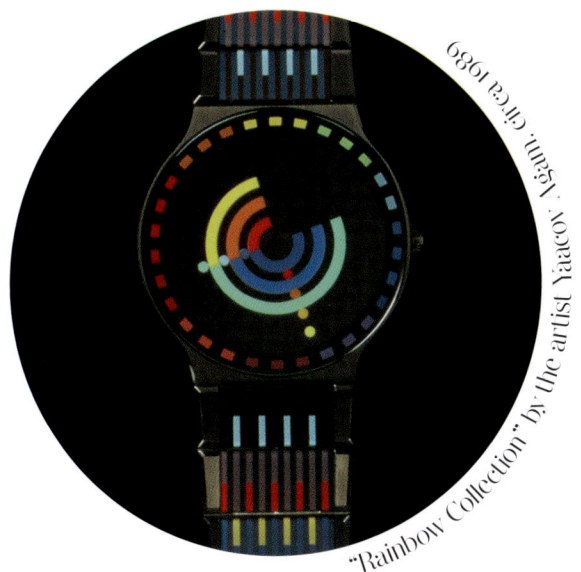
"Rainbow Collection" by the artist Yaacov Agam, circa 1989

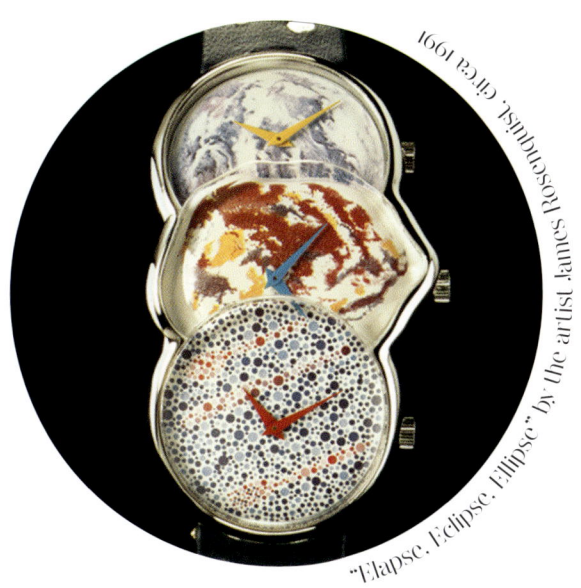
"Elapse, Eclipse, Ellipse" by the artist James Rosenquist, circa 1991

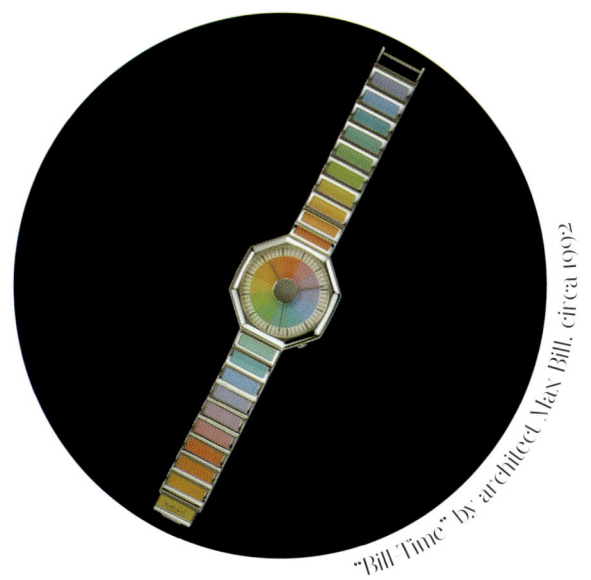
"Bill Time" by architect Max Bill, circa 1992

Perhaps more so than any other brand, Movado (whose name aptly means "always in motion") has long been committed to the arts. It was in 1988 that the brand launched The Movado Artists' Series of limited edition timepieces designed by world-renowned artists. Gedalio Grinberg, then owner of the brand, commissioned the artists to use the watch as an art medium to portray their philosophy of time. Just about a dozen or so Movado Artists' Series Limited Editions have been designed so far by some of the most influential artists of the 20th century.

ENCHANTING ENAMELS

Easily one of the most magical and alluring arts transformed in miniature onto watches is that of enamel work. This is such a time-honored traditional craft that just a few of the finest artisans in the world are capable of bringing watch dials to life with extraordinary vibrant scenes, lush flowers, lifelike animals and detailed animations. Most of these artisans have been practicing their craft since they were young teens. They spend hours and hours every day painting a single dial — sometimes with a brush with just one animal hair to achieve the tiny wisps that comprise a bird's feather or a person's eyelash. // To complete a single enamel watch dial requires hundreds of hours of tedious painting. What's more, a dial undergoes dozens of firings in an 800-degree kiln — subjecting it to damage or breakage, wherein the enameller must throw it away and start all over again. In enameling, colors cannot be mixed. Instead, the artist must paint a color and then fire it, before starting another color. In the hot firings, sometimes the colors change and lighten. To achieve a softer, more translucent hue, fewer layers are added, but if a rich, dark color is desired, the enameller must paint and fire coat after coat. Often a watch dial can have as many as twenty paintings (and firings). There are several different types of watch dial grand feu enameling, including cloisonné, where strips of gold are laid on the dial to form the outline of the piece to be painted, and then the enamel is filled in. Champlevé, flinqué and spangle (featuring the placing of gold foil decorations sunken into the enamel) are also important techniques. Then, of course, there is free-hand enameling — leaving the artist to create at will with unbridled passion.

Part of Vacheron Constantin's Métiers d'Art – Chagall & l'Opera de Paris collection, this watch is the first in the series called "Tribute to Famous Composers." Since 1755, Vacheron Constantin has been perpetuating the fine arts of watchmaking and its related traditions. This line of superb grand feu enamel watches consists of 15 one-of-a-kind models in tribute to the greatest composers – those who inspired Marc Chagall for his monumental fresco painting adorning the ceiling of the Garnier Opera House. This watch was unveiled in 2010, with the other 14 watches in the line being created over the next couple of years, each dedicated to one of the composers appearing in Chagall's work. Chagall's work spans over 200 square meters – this dial is 31.5mm – an amazing feat.

"The dignity of the artist lies in his duty of keeping the sense of wonder in the world."

— Marc Chagall

The Artist's Take On Time

Just as a museum curator selects and organizes an art show, some watch brands seek the finest artists in the world to paint their watch dial canvases. Other times, the artist seeks the medium.

In free-hand miniature painting — where the enamel is mixed with oil and the artist paints freehand — as in this Vacheron Constantin watch — everything is left to the artist's fine hand, amazing skill and grand imagination.

Some artisans will create a blend of the different mediums and develop their own signature look or color. Hundreds of hours go into the making of each dial. With enamel work, every watch is a unique piece, individually painted and created with passion and love.

At the end of the 19th century Bohemian Alphonse Mucha, new to Paris, wrote to French stage actress Sarah Bernhardt requesting an introduction to the important figures of that metropolis. His posters for her were introduction enough, and she wrote back. Their friendship inspired many of his works — including jewelry, soap labels, mosaic panels for swimming pools, textiles, calendars, letterheads, ads and stage art. The great jeweler Fouquet translated some of Mucha's finest pieces (such as a snake bracelet and ring) for Sarah. His line became known as the line of the new art (Art Nouveau), and he became the most celebrated artist in Paris. He formed a commercial love alliance with Sarah Bernhardt, when he designed the poster for the great play Gismonda — perhaps the greatest theatre poster ever created — and thereafter designed all her posters, together with costumes, sets and personal knick-knacks. But Mucha needed the heady airs of Parisian cosmopolitanism.

To celebrate Mucha's art, Corum recreated on watch dials several of the famed Mucha posters made for Sarah Bernhardt. The Classical Mucha collection is a limited edition of four dials — 24 pieces of each — hand-painted on mother of pearl.

SCULPTURED BEAUTY

In real-life art and in the most celebrated sculptured watches, artists work in many mediums. Sculptured timepieces have been a part of watch history for centuries. Designers first engraved in the metal and then later began adding adornments such as inlaid precious stones — carved and cut to add depth and dimension. Today's artisans have brought this concept to compelling new levels — constantly pushing the creative envelopes to offer astounding masterpieces. The work is slow going and requires an incredibly steady hand and much patience. However, like any other artist, these professionals are passionate about their creations and each dial — unique unto itself — is its own canvas.

This Van Cleef & Arpels "Makis" Haute Joaillerie Timepiece is crafted in white gold and set with diamonds, blue and pink sapphires, and onyx. The dial features mother of pearl inlays. It houses a Swiss quartz movement and is a unique piece.

© Attitudes Photo/Rémi DERYCKE (Courtesy of Van Cleef & Arpels)

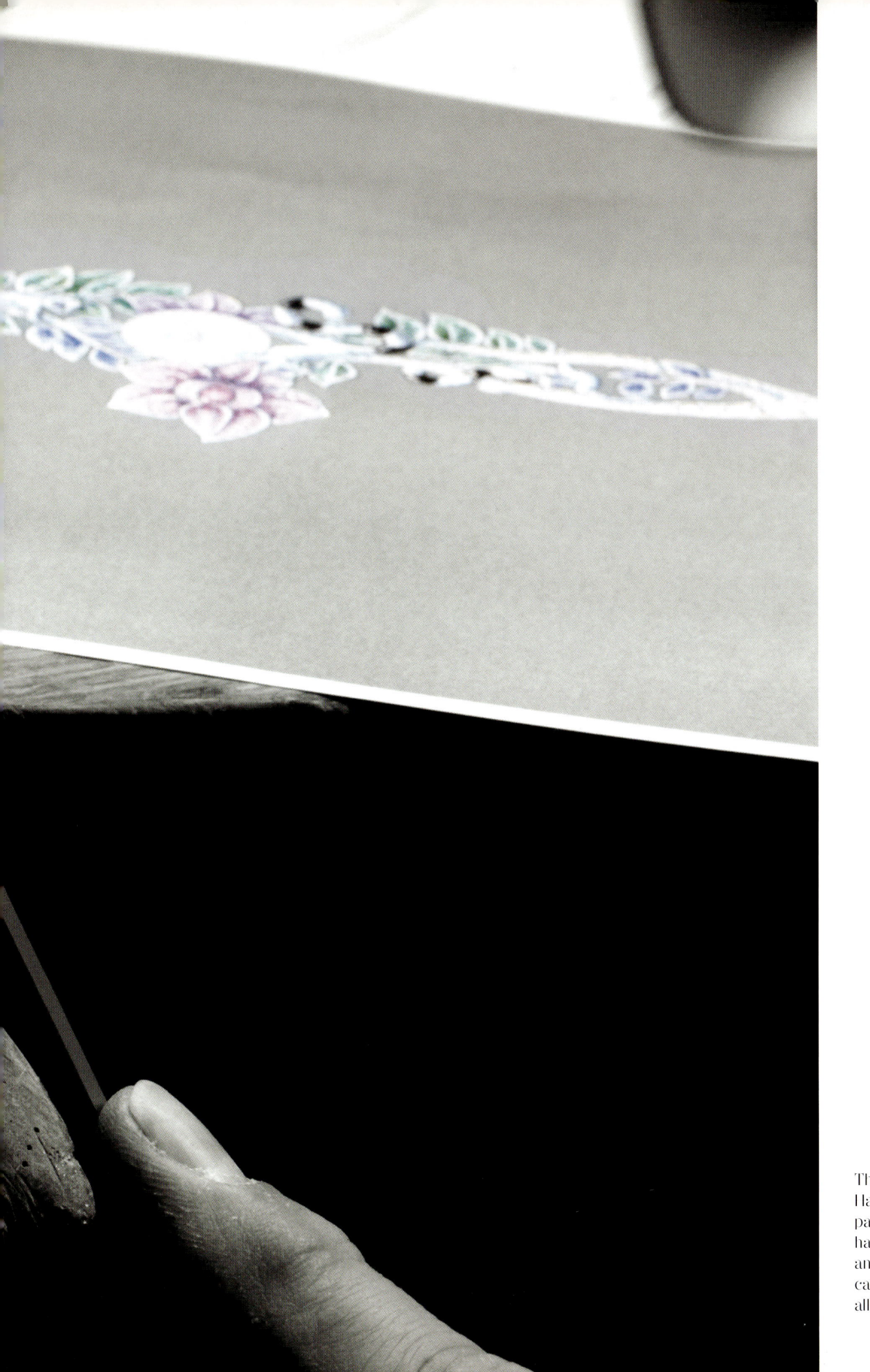

The making of the Van Cleef & Arpels Haute Joaillerie Timepiece is a long and painstaking process done entirely by hand. After all sketches are complete and colors selected, the gold is hand carved for the meticulous gem setting — all of which can take hundreds of hours.

© Attitudes Photo/Rémi DERYCKE (Courtesy of Van Cleef & Arpels)

© Attitudes Photo/Rémi DERYCKE (Courtesy of Van Cleef & Arpels)

MYSTICAL FIRE-BREATHING DRAGONS

Dragons — those wonderful mythical beasts who breathe fire and command absolute attention — these are the creatures that fascinate all. It is no wonder that tales, novels and history are wrought with these almost majestic animals and the men who wish to slay them. Perhaps that is why so many artists in time — including today's finest watchmakers — slay them in their own ways, bringing their beauty and mythical wonder into new dimensions on the dials of timepieces. Whether in sculpture or enamel, in paint, gold or other form, the dragon-designed watch goes down in history as one of the most storied and individual artisan creations in time.

This Corum Classical Pearl Dragon watch is a stunning hand engraved and painted dragon via a patent pending process on a pearl dial created in an extremely limited edition. It celebrates the ancient Chinese symbol of power and fortune. Housing a mechanical movement, the bezel is set with diamond brilliants.

"For the woman identifying with a solid symbol of strength or luck, the Imperial Dragon speaks to those special places within you, stoking the fires of your heart."

— Thierry Oulevay, Co-Founder, Jean Dunand Pièces Uniques

The Jean Dunand Tourbillon Orbital Dragon watch houses an exceptional one-minute flying tourbillon that orbits the dial once per hour on a patented revolving movement. This Pièces Uniques watch is exceptionally hand enameled with the dragon motif on a black onyx dial.

"Traditions evolve, but never disappear." — Zanetti

Zannetti Lady Regent Legendary Dragon watch with Regent Dragon sculpted in 18-karat gold: its sinuous body dominates the scene, contrasting with the enamel dial. The Italian-made timepiece houses a Swiss automatic movement and the 18-karat gold case is set with brilliant-cut diamonds.

"Challenge is a dragon with a gift in its mouth — tame the dragon and the gift is yours."

— Noela Evans

A passion and inspiration for nature inspired Carrera y Carrera to create this elaborate collection of jewelry watches that features an exclusive concept of bezel inserts and jewelry covers, offering a sophisticated yet baroque look. This Círculos de Fuego watch features an interchangeable bezel to wear the watch plain or with the alluring dragon in 18-karat gold filigree work.

Piaget Limelight Garden Party watch, handcrafted in 18 karat white gold, and featuring a black dial with white gold rotating birds and leaves set with brilliant-cut diamonds on a black satin strap with ardillon buckle. A total of 165 brilliant-cut diamonds were used.

TIME IN FLIGHT

One of the watchmaker's most treasured subjects is nature because of the richness of its colors and intricacies. Nature has always been man's best inspiration. Without it, time would stand still. We would most likely falter. Our best concepts derive from a moment of solitude in nature, or from inspiration taken when we admire a spectacular field of flowers or bird in flight. Nature is an artist. So, too, are some of the finest watchmakers who try to recreate nature's beauty on our wrists. // Birds are particularly stunning thanks to their many wonderful poses, and the freedom they inspire in each and every one of us. They spread their wings and take flight — soaring through the sky like a moment soars through time. Or they rest a moment, perched somewhere as they enjoy the sun and the flowers, and sometimes let the human eye enjoy them. Enamellers and other artisans, including gemsetters, capture those feathered pauses, those open winged moments of flight and those perfect landings in all their glory — giving us not just breathtaking works of art, but also the ability to contemplate time and the beauty of flight.

"The very idea of a bird is a symbol and a suggestion to the poet. A bird seems to be at the top of the scale, so vehement and intense his life... The beautiful vagabonds, endowed with every grace, masters of all climes, and knowing no bounds — how many human aspirations are realised in their free, holiday-lives — and how many suggestions to the poet in their flight and song!"

— John Burroughs
Birds and Poets, 1887

"If I had to choose, I would rather have birds than airplanes." — Charles Lindbergh

This elegant Jaeger-LeCoultre Master Lady Tourbillon is a parade of talent with mother of pearl, aventurine, enamel and graceful realistic bird pose protecting the tourbillon escapement.

(Left:) Drawing of the Santos 100 watch with the hummingbird motif (© Ines Dieleman / Cartier)

(Right:) With its wings aflutter, this Cartier Santos 100 Hummingbird motif watch is a delight to behold. It is a magical blend of mother-of-pearl marquetry, champleve grand feu enamel and magnificent gemsetting using 20 pink sapphires on the dial and 216 diamonds that combine to depict the world's smallest bird in all its natural splendor. This feminine creation is composed of ten pieces of mother of pearl and nine colors of enamel. (© Ines Dieleman / Cartier)

Q&A

with Cristina Wendt-Thévenaz, Delaneau

What does time mean to you?

Time means constant change, whether we like it or not. I do not view time... I feel it.

How does time influence us?

The only moment really important is now... it reflects our past and influences our tomorrow.

How does a woman view time differently than a man?

A woman does not like to measure time; she would love to control it.

How does a watch really reflect time?

A watch is a beautiful reminder of endless flow. The fact that you have to provide the "energy" for this mechanism to help us grasp time is a wonderful way to keep in touch with something that we cannot really control.

From DeLaneau, these Atame enamel bird watches are from the brand's Pair collection, distinguished by its elegant, perfectly balanced rectangular form and unique case-to-bracelet lugs. Atame means "attach me" in Spanish. The dials are entirely hand engraved and enameled and the watches are powered by mechanical movements.

This Van Cleef & Arpels "Blue Throated" timepiece is from the "Colibri" Extraordinary Dials Timepiece collection. The hand-enameled hummingbird is drawing nectar from a rubellite flower. The watch is set with 103 diamonds totaling 2.7 carats in 18-karat white gold.

"Women appreciate the poetic dimension;

...they enjoy the movement of the watch, and the miniature reality. They like to play with it and show it off; they have a real appreciation for the playful dimension."

— Nicolas Bos
President and CEO, the Americas,
Worldwide Creative Director,
Van Cleef & Arpels

"I started setting diamonds in watches when I was young, before I went to college. Then I tried to do other things, but I missed setting the diamonds. You are never bored as a gemsetter. You always find something new, new pieces to work on, new ideas, and new designs. You develop your own skills and leave your own signature on these watches, and you hope that it is something every woman wants to wear. I am passionate about it. I cannot explain what I am thinking as I am setting these wonderful stones into beautiful pieces...

"and I know that one day it will be worn by someone who loves it, that it could be part of history, that it will be in a picture."

— Arthur Guessian,
Gem setter, Van Cleef & Arpels

From Van Cleef & Arpels, this "Calliope" timepiece from the "Colibri" Extraordinary Dials Timepiece collection is set with 172 diamonds totaling 2.8 carats set in 18-Karat white gold. It features a gold-sculpted and translucent enamel dial.

"One should either be a work of art...

(Above:) This Boucheron "Wings of Desire" watch features a pink sapphire flamingo that stands on a mother of pearl dial while a school of golden fish swims about its ankles, and green tsavorite pussy willows billow around it.

(Right:) DeLaneau enamel and gem-set hummingbird watch

"...or wear a work of art."

— Oscar Wilde

Q&A

with Caroline Gruosi-Scheufele, Co-President and Artistic Designer, Chopard

If you had an extra hour of time each day, what would you do with it?

Time is the essence of everything. If I had an extra hour of time each day, I would turn my phone off... I would spend it with my loved ones.

What does time mean to you?

Time is infinite, yet the human existence is not, therefore time is extremely precious and should be handled with care.

Chopard High Jewelry Owl Watch

conceived of in collaboration with the legendary Boucheron, the MB&F JWLRYMachine is predicated on the HM3 platform – housing a superb mechanical movement that made its debut first as a magnificent new timepiece, then interpreted as a frog and now as an avant-garde bejeweled winged beauty. "For me, when creating a watch there is always a human story," MB&F founder Max Busser explains. "This one was designed with Boucheron, who creates amazing three-dimensional jewelry and I design 3D watches and we knew we could do something very special together. When we first started we decided that the main goal was to build something that would make people smile and be shocking. Women see this watch as a work of art, and that is what is was meant to be: a JWLRYMachine that is art, simple and enjoyable." Solid gold renditions include amethyst and violet sapphires with diamonds, or 18-karat rose gold with pink sapphires, pink tourmalines, rose quartz and diamonds. In this watch, the rotor lies beneath the center faceted stone of the owl – so it looks as though the bird's feathers are ruffling or its heart is beating with every turn of the rotor. Naturally, it is created in extremely limited numbers.

Urushi is the ancestral Japanese art of lacquering using a varnish created from the sap of a lacquer tree — or Urushi — that is mainly found in Japan and China. Harvesting the resin of the tree is done in the smallest of quantities only once a year, and then the resin sits for three to five years before it is treated to make a honey-textured lacquer. Maki-e is a specific technique that incorporates the sprinkling of gold dust via bamboo tubes and small brushes made of rat's hair to create incredibly fine lines. These dials, created by Urushi Masters, including Kiichiro Masumura, are unique works of timeless art.

Capturing the quintessential Japanese art form, along with Chopard's L.U.C watchmaking prowess, the Chopard XP Urushi houses an ultra-thin self-winding 96HM movement inside. Nine different dial models are created in gold and lacquered with varnish made from the sap of the Urushi tree, hand painted by Kiichiro Masumura.

"The pride of the peacock is the glory of God."

— William Blake

LeVian Time® Peacock from the "Into the Wild" collection is set with five carats of diamonds, tsavorites and blue sapphires and accented with a stingray strap.

Known for centuries as one of the great innovators in the jewelry world, the great House of Boucheron has teamed with Girard-Perregaux to bring timepieces of haute horology and haute joaillerie to the surface. This one-of-a-kind Hera Tourbillon is an amazing example of beauty and awe that demonstrates the French brand's amazing savoir-faire. Beating at the heart of this peacock timepiece is a custom-made Three Gold Bridges Tourbillon from Girard-Perregaux. The mechanism is set with two different shades of green tourmalines, along with diamonds, to create the body of the peacock. A ring of 52 round-cut diamonds encircles the watch case to form the body of the alluring sculpted peacock, whose feathers are set en-tremblant — meaning they move with wearer's movements. The Peacock consists of more than 35 carats of diamonds, Paraiba tourmalines and sapphires. In fact, the entire bracelet is set with 271 diamonds, 868 blue sapphires, 426 purple sapphires, 310 tourmaline Paraibas and one blue cabochon sapphire. The magnificent watch is expected to retail for just under one million dollars.

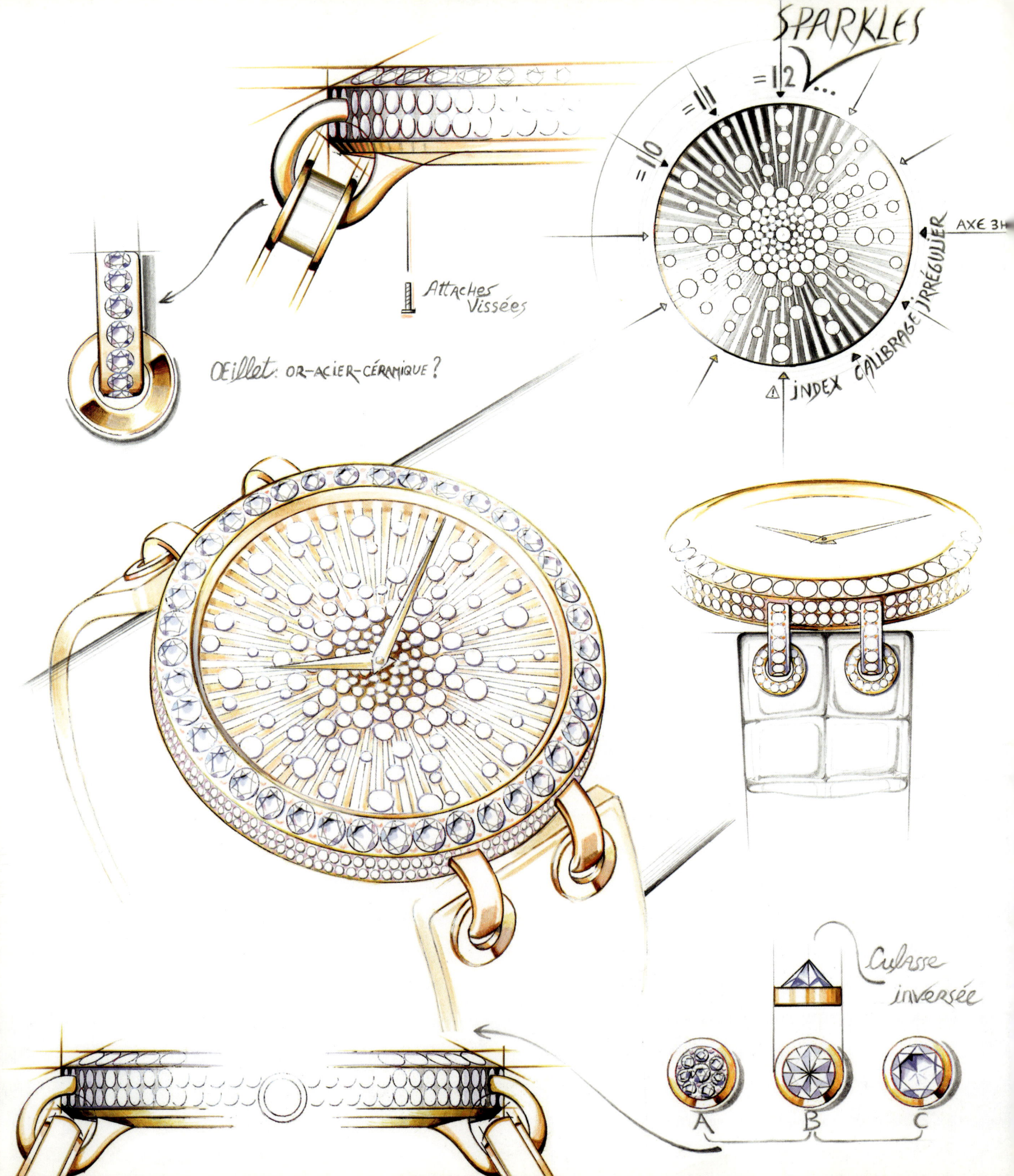

CHAPTER THREE ARCHITECTURAL INFLUENCES

Perhaps more so than anything else, the watch is a reflection of progress: time on the move. While artisans can manipulate the dial with paintings and centuries old crafts, watch designers and watchmakers build architectural delights. Indeed, watches are inspired by structure, by architectural lines. Many of today's finest pieces take their form — from case, to bracelets, to lugs — from the architecture that surrounds us. It can be a building or a city in the world that lends its way to a dream. Other designers recall the past and are enlivened by bygone eras such as the lively Art Deco and Art Nouveau spirits. Not only does architecture exist on the outside structure of the watch, but also, and perhaps even more so, on the inside. The extreme precision and mechanics of a finely tuned, highly calibrated engine — hundreds of components working together in a technical mechanical movement — is like an orchestra of art — an architectural wonder.

Sketches of the incredible Chopard Haute Joaillerie Xtravaganza watch

TIME AROUND THE WORLD

Artists in all walks of life, from painting to photography and sculpture, are continually awed by some of the world's most alluring cities and their important architectural contributions to history. It is often the most expressionistic concrete or iron buildings that have given inspiration to so many masterpieces — when these structures are works of art in and of themselves. How could a watchmaker with a vision pass by these same architectural wonders, or through the streets of lights and music, and not be driven to set their themes to time?

Jacob & Co. Five Time Zone "The World Is Yours" watch offers time in five zones. The dial is set with 2.50 carats of diamonds.

The Tiret "Second Chance" Chrysler Building watch offers double time zones, double chronographs and double date windows, along with a tribute to some of the finest architecture in the world.

"Architecture should speak of its time and place but yearn for timelessness."

— Frank Gehry

"Women appreciate style, mechanics and structural beauty. Today's women know how to be their own person."

— Jacob Arabo, Founder, Jacob & Co.

Jacob & Co. creates cutting-edge timepieces that combine bold statements of individuality and personality. (Left:) This Limited Edition stainless steel Moscow Watch is a three-dimensional dial with 50 white pavé round diamonds and 132 rubies, and multi-colored enamel buildings. (Right:) The Jacob & Co. Limited Edition Manhattan Watch is a three-dimensional piece featuring 162 Round Cut Diamonds and 20 baguettes. It is crafted in 18-karat white gold and set with an additional 498 round cut diamonds.

ART NOUVEAU & ART DECO

The Art Nouveau era of stylized art and architecture that ushered in the 20th century had its effect on timepieces and was especially enthralling for women's watches. This was a time of utter elegance, of bows and tassels, of flowers and vines. The finest watchmakers in the world found ways to interpret these elements onto timepieces that reflected the expression and artistic distinction of the time. When this era of flowing organic curves gave way to the more classical, structured linear symmetry of the Art Deco design styles of the heady roaring twenties, some of the most extraordinary watches emerged. Many designers looked to the luxury steam liners of the day and the most superb buildings in Art Deco style to influence their watches — offering timepieces of true distinction. The watches of the Art Deco era, which spanned nearly 20 years, were diverse in styling, but were always elegant pieces that utilized either unusual shapes, stunning combinations of the Earth's most precious stones, or beautiful patterns that led to a geometric delight. Today, many watch brands continue to create these architecturally perfect Art Deco-inspired designs — recognizing that true art is as timeless as the watch.

This stunning Boucheron Ma Jolie watch is crafted in white gold set with diamonds and a mother-of-pearl dial set with diamonds. The entire bracelet and corollas are in white gold set with diamonds.

Art Deco Chrysler Building elevator door, photo courtesy of Tishman Speyer

At Chopard, Art Deco inspiration was interpreted through construction techniques and design was at the forefront. This intense attention to innovation, high craftsmanship and detail continues today. This stunning fan-shaped melded 18 karat gold watch utilizes the Happy Diamonds free-floating diamonds principle for alluring appeal.

Carl F. Bucherer created an extraordinary watch to pay tribute to Wilhelmina "Mimi" Bucherer Heeb, who was born in 1899 and whose entrepreneurial and adventuresome spirit was the driving force behind her husband Carl. She is credited with bringing one of the first wristwatches created especially for ladies — a diamond adorned work of Art Deco — to Chile with her from Switzerland in the early 1920's. The Tribute to Mimi watch pays homage to this extraordinary Swiss woman.

The Tribute to Mimi watch by Carl F. Bucherer houses a vintage Ebauches SA 715 caliber mechanical hand-wound movement that offers hours, minutes and small seconds. The elegant 18-karat white gold watch is set with 108 diamonds on the case weighing 1.45 carats and another 420 diamonds on the bracelet weighing 1.7 carats. Only 70 were ever created.

Because of its beauty and rarity, the ruby has been considered throughout history as the mother of all precious stones. The ruby was long considered to have divine powers and to offer energy, love and spiritual strength. It has been used over the centuries as a talisman, a symbol of power and bravery and as a marker of wealth. During the Gothic era, the color red, along with black and gold, was especially important in churches. In that respect, the Carl F. Bucherer Alacria Diva Gothic watch displays a typical feature of that period with its curved profile — it is a jewel for the self-assured woman. Crafted in 18-karat white gold, the case is with 54 baguette-cut rubies weighing 5.0 carats, 157 rubies weighing one carat, and 82 full-cut Top Wesselton VVS diamonds weighing 1.3 carats. The dial is set with 74 diamonds weighing 0.7 carats and outlined with onyx. The watch, created in a limited edition of just 25 pieces, is accented with a black stingray strap. The watch is set with 211 rubies and 169 diamonds FC. Just 25 pieces will ever be made.

"Time is part of our lives; apart from being enjoyable, it can also be fashionable and stylish. I love that the watch can make an artistic statement of highest craftsmanship."

– Michelle Yeoh, Actress, Design Consultant to Richard Mille

This Richard Mille RM 007 Ladies Automatic is crafted in 18-karat gold and magnificently set with 62 pink sapphires, 42 pink sapphire baguettes, and 104 pink diamonds totaling 5.30 carats.

At Van Cleef & Arpels, gemsetting is an essential part of the creation and is decided upon very early in the creative process. The highly precise operation is influenced by the personal style of each artisan. This Le Nôtre watch features captivating diamonds and emeralds in all their glory.

"It can take three months to finish setting a single watch, but I don't think of the time because with this work you are putting part of your life into it.

Each piece you touch, you feel;

and when you have finished, it is a great feeling to know you have created something people will love."

— Arthur Guessian, Gemsetter,
Van Cleef & Arpels

This Damiani Belle Époque Masterpiece 2 is crafted in 18-karat white gold and features an internal bezel set with 24 baguette-cut diamonds and 12 sapphires, and a rotating external bezel set 56 baguette-cut diamonds with an "extra-sized" sapphire.

"For today's woman the watch is no longer considered an accessory — it is a must that women utilize to complement their look and organize their schedule. The fast pace of her lifestyle keeps her active and busy in many fronts, at times juggling the responsibilities of family and career; while at the same time aiming at looking her best."

— Silvia Damiani, Vice President, The Damiani Group

Damiani Belle Epoch haute joaillerie mystery watch set with baguette cut diamonds and sapphires.

"I chose the diamond because its density represents the greatest value for the smallest size."

— Coco Chanel

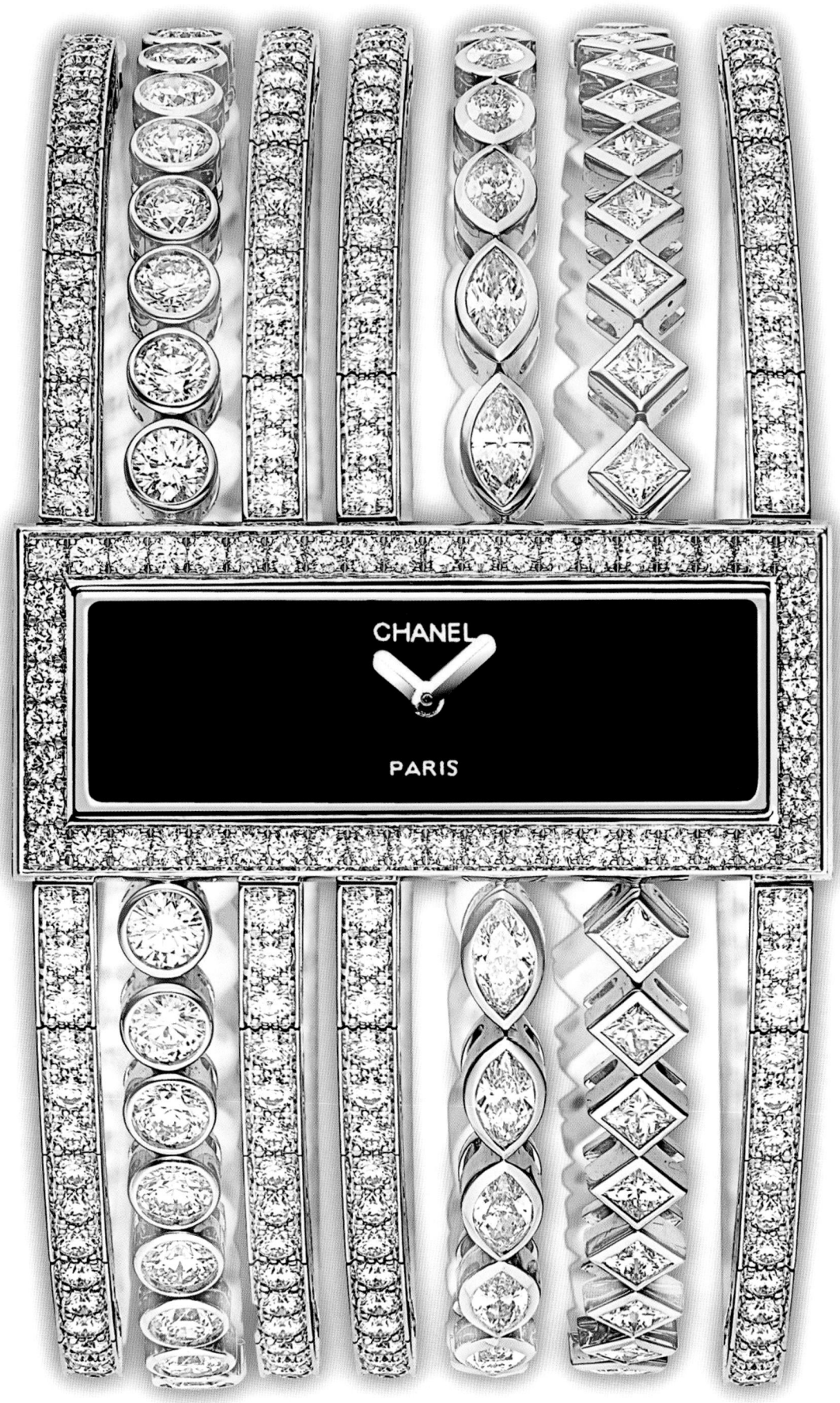

Chanel Fil de Diamants watch

This Harry Winston Lattice watch is part of the Jewels That Tell Time haute joaillerie collection of diamond watches that demonstrate the legendary diamond setting techniques of this great House. Capturing the Harry Winston signature style, the Lattice features differently shaped diamonds set at slightly different angles. It has 592 internally flawless Harry Winston diamonds ranging from D through F and weighing a total of 9.57 carats.

"For today's woman, time is a valuable gift. Her watch is not merely a tool to view the passing hours, but an instrument to measure each meaningful moment. Timepieces for women should reflect this philosophy in many ways, such as the rare beauty of a perfectly set diamond for the Jewel that Tells Time or in the purity of design."

— Sandrine de Laage, Art Director & VP of Design, Harry Winston

Bulgari Via Dei Condotti watches are crafted in 18-karat white gold and diamonds in unusual shapes.

MECHANICS OF TIME

Within every watch there beats a soul — the tiny movement that keeps pace with a woman's every move. Today's most prestigious brands recognize that women of taste have a true appreciation for architecture and beauty — for a mechanical movement to power the art form on their wrist. These mechanical calibers are typically hand assembled and often hand finished. Many comprise hundreds of miniscule parts — each put together one by one — to ultimately rotate, coil and vibrate in perfect harmony like an orchestra of time. Generally the final movements, sometimes visible via a transparent sapphire caseback or some small dial aperture, are finely, eloquently finished with engraving and detailing that defines the watch as a special hallmark. // It can take weeks — even months — for the most complicated of movements to be assembled before being fit into a watch case of equal distinction. Absolute concentration, extreme patience and fervent passion — along with years of experience — are what ultimately combine to form a superb mechanical watch. Many of the most exclusive pieces are created as limited editions or one-of-a-kinds and can cost tens of thousands of dollars — even hundreds of thousands — just because of the movement within. Sometimes there is even a waiting list for these mechanical marvels — but, as every woman knows, good things are worth waiting for.

From Armin Strom, this Skeleton Square Lady watch is hand made as part of a small, strictly limited edition. It houses a rare 1970's Alpina movement that has been recalibrated and built up by Armin Strom as the Caliber ASL07, hand wound fully skeletonized and hand engraved. The 18 karat white gold case and bezel are rectangular octagonal in shape and set with 33 diamonds.

"Nothing valuable can be lost by taking time." — Abraham Lincoln

Girard-Perregaux Cat's Eye Annual and
Zodiacal Calendar

"We need not feel ashamed of flirting with the zodiac. The zodiac is well worth flirting with."

— D. H. Lawrence.

This Girard-Perregaux Cat's Eye Annual and Zodiac Calendar watch was two years in the development stages. Created expressly for women, the timepiece houses 287 tiny parts in its complex movement. The making of the dial is also one of the most critical factors, as it consists of two main plates, an 18-karat gold base plate and a mother-of-pearl upper late. Up to 45 different operations go in to the making of a single dial, with steps such as circular graining, zodiac sculpting, marquetry of the mother-of-pearl and exacting cutting and fitting. The case of the timepiece is pavé set with 315 diamonds for a shimmering finish. Because of the complexity of the making of this creation, just about 200 pieces are made per year. The watch offers an annual calendar with date, as well as the signs of the zodiac.

A. Lange & Söhne may well be the only watchmaker in the world that employs more women than men in its upper echelons of artisanry. In fact, 56 percent of this brand's employees are female... watchmakers, engravers and top artisans. Bravo.

"The love and passion for watches should really be in your blood. Sometimes I think that my heart is ticking, not beating."

— Heike Ahrendt, Product Manager;
A. Lange & Söhne

"I believe that women are more sensitive, with a finer touch, when it comes to engraving. They often work an engraving more delicately and in a more filigree way than men. And at times we are more persevering, always trying to go further by adding a little ornament here and there or putting in an additional line in order to make the picture look even more perfect."

— Simone Rauchfuß,
Engraver, A. Lange & Söhne

This A. Lange & Söhne Little Lange 1 Soirée is an enchanting gold and mother-of-pearl jeweled work of art hailing from the German town of Glashütte. Beating at the heart of this lovely piece is the Lange manufacture caliber L.901.4 manually wound Lange 1 with an amazing 365 individual parts in it, including 53 jewels that keep it beating at 21,600 semi-oscillations per hour. This gem of a watch offers a patented outsized date indication in addition to 72 hours of power reserve. The case is set with 72 Top Wesselton diamonds weighing 1.1 carats.

IWC's alluring Da Vinci watch houses an automatic movement and offers date indication. It is elegantly crafted in 18-karat rose gold with a sunray guilloched dial.

"Time stays long enough for anyone who uses it."

— Leonardo da Vinci

This Oyster Perpetual Lady Datejust is crafted in 18-karat Rolesor gold and features a stunning dial of Gold Crystals as an exquisite homage to femininity. It is created using an exclusive process that magnifies the natural crystal structure of the gold and gleams with extraordinary reflections. The Jubilee model features a bezel set with 46 brilliant-cut diamonds. The stunning watch houses a Superlative Chronometer Officially certified COSC 2236 Manufacture Rolex movement that is water resistant to 330 feet.

"Perfection is attained by slow degrees; she requires the hand of time."

— Voltaire

This Patek Philippe Ladies Grand Complication, Ref. 7059R houses the world's thinnest Split Seconds Monopusher Chronograph with column-wheel control. Inspired by an original vintage officer-style case and redesigned in a feminine matte finish cream colored dial, the manually wound column wheel chronograph movement offers a 60-minute counter and a seconds subdial. The watch is crafted in 18-karat 4N rose gold, and the bezel is set with a double row of 150 diamonds. It is valued at nearly half a million dollars.

"Women used to like brand names more than the product, but today's woman, fortunately, is so much more attuned to the product. She is more interested in what's inside the watch — and it's even better when she can see that mechanical movement through the dial — see the heart of the watch beating."

— Peter Stas, Founder,
Frédérique Constant

> "Time is the greatest innovator."
>
> — Francis Bacon

The Corum Golden Bridge Lady takes its inspiration from the first Golden Bridge watches that were designed in 1980 and have since evolved into spectacular, elegant baguette shapes for women. This more feminine version is a splendid reinterpretation in mechanical magnificence. The sapphire case enables one to view the delicately engraved gear trains in motion. The watch is a contemporary perfectly proportioned work of art and craftsmanship.

"Women's watches are influenced by all of these issues (art, architecture, style), indeed, perhaps more so than men's watches. Of course all watches center on an emotional connection, however women's watches require a clear thread within the emotional world that is represented. Even with the most artistic jewelry watch, inspired by artistic imagery, women want to know what is behind the imagery and what the emotional content within it is. Whether it is a free artistic or architecturally inspired piece, I am always intrigued to see that women want to know what is at the heart of the idea."

— Richard Mille

Assembled barrel bridge, assembled center bridge, and lower barrel bridge used in the Richard Mille RM 002 V2 and RM 003 V2

Q&A

with Guillame Tetu, Managing Director & Co-founder, Hautlence

What do you think women want in a watch?

One would think that women need greater precision — and not that precision is not a quality found in an Hautlence timepiece — but one can find so much more, such as elegance, balance, personality and above all an attention to detail and harmony. A woman wearing a Hautlence watch stands out as a person of character who has definite tastes and has made choices that don't necessarily fit the common standards; she often is the driving force and the leader in the room.

Do you think women view time differently than men?

Women are much better customers than men! In French we speak about the 'contenu' and the 'contenant' or, if you wish, the inside and the outside. Our women customers have the ability to understand the broader beauty and value of our timepieces while really visualizing the inside and outside as one complete work of art. The daily communion between wearer and machine in the form of rewinding the mechanism of our exclusively mechanical timepieces for a mere 22 to 28 seconds becomes a sensuous marriage where women more than anyone else can sense the intricacies of the finesse of the open connecting rods and jumping wheels associated with our instruments.

Why did you design a mechanical watch for women?

Ever since the 1920s, women have come out of the stereotypes and boundaries established for them by men and society. Smoking, dancing, voting, driving and so many other pleasures associated with material wealth started to be accessible to them. In examining the demand we have from our consumers, we realize the extreme high level of sophistication and taste. The requests are often exquisite. We have demands where the wealth of materials is as much important and rare as the precision of the timepiece and the visibility of the various components of the caliber employed by our watchmakers.

The Hautlence HL.co3 watch with violet mother of pearl layered dial is crafted in gold and offers jumping hour and retrograde minutes on a minute track with volumetric minute figures. The mechanical manual wind movement consists of 24 jewels and beats at 21,800 vibrations per hour. It is created in a limited edition of just 88 pieces.

"The mother art is architecture. Without an architecture of our own we have no soul of our own civilization."

— Frank Lloyd Wright

This Hautlence HL.c04 watch is created in 18 karat gold with 102 diamonds on the case. It is a jump hour, retrograde minute, manual-wind mechanical movement watch.

"The thing you can never buy is time. We can never get it back, so how we spend it and use it is very important."

— Nicole Kidman,
Omega Ambassadress

Omega Watch Company's Ladymatic watch series was two years in the making. It is a dramatic mechanical watch line with fashion-forward visionary appeal. It houses a state-of-the-art Co-Axial movement and high-tech ceramic materials wrapped into its 18-karat gold case. The collection's name stems from the company's archives, dating back to a series originally launched in 1955, but that is the only thing old about the line. Its design inspiration — taken from the movement of the ocean's waves — is all new. Each watch is a 34mm COSC-certified chronometer. The Co-Axial movement features an exclusive Omega Si 14 Silicon balance spring — a rare feat in watchmaking. The watch offers a sapphire caseback to view the extraordinary caliber, and features an alluring patented three-row bracelet with asymmetrical links.

This Peter Speake-Marin watch from his Sea & Stone Collection has a parquetry dial designed with 36 pieces of mother-of-pearl and onyx.

"Everybody is limited to time and every day is a gift. My designs come from my soul, from the way I see life. They represent who I am as a watchmaker and as a person."

— Peter Speake-Marin, Watchmaker

Women are like watches...

(Above:) Ulysse Nardin Executive Lady Dual Time watch houses the brand's UN-24 dual time automatic movement. The second time zone hour is read digitally via the aperture on the left of the dial.

(Right:) Parmigiani Fleurier Pershing chronograph watch with mechanical movement and chronograph functions for count-down timing of events.

they tick perfectly even when you're not looking...

Recently, many top brands have been working with high-tech materials such as ceramics for luxury timepieces. The J12 Intense Black watch from Chanel's J12 collection is comprised of an 18-karat white gold case and stunning baguette cut black high-tech ceramic pieces. Chanel collaborated with Swiss watchmaker Audemars Piguet for the self-winding three-hand Haute Horlogerie movement.

"Fashion is not something that exists in dresses only."

Fashion is in the sky, in the street, fashion has to do with ideas, the way we live, what is happening."

— Coco Chanel

Applying diamond cutting techniques. Chanel craftsmen from the prestigious atelier of La Chaux-de-Fonds spent 350 hours faceting 724 miniature black ceramic pieces into baguette cuts before individually setting each tile into a white gold frame.

Image courtesy of Baume & Mercier

CHAPTER FOUR
DESIGN INSPIRATIONS

There is a certain duplicity, even multiplicity, about watches. They are a statement of style, design and fashion — an expression of life and changing times. Many watch designers say they are inspired in their creations by life around them — by clothes, color, nature and even by life's secrets. They create watches that are resounding reflections of reality — and for this reason, art on our wrists constantly evolves, shaping our view of the ebb and flow of time, and influencing the way the future views the past. Other watch creators say they want their timepieces to be ethereal pieces, dreams, fleeting moments that are the antithesis of reality. For them, there are secrets in time, hidden watch dials, unusual shapes and time in all dimensions. This is what life is made of... diversity, choice, innovation, moments of individuality and a selection that shapes time forever.

This alluring Piaget Miss Protocole Lilac strapped watch is a wonderful enchantment thanks to its whimsical feathered bracelet.

ICONS IN FASHION

In history, fashion influences most aspects of a woman's life. One cannot say fashion defines the woman, but rather, the woman defines the fashion — wearing a certain designer name, a key look, a signature pattern. Thus, it comes as no surprise that by the latter portion of the 20th century, the finest designer and haute couture names were recognizing that they should spread their proverbial wings into the world of watches. While some of them had already built an occasional watch here and there along the way, others delved headstrong — like a woman — into the category. Today, designer watches make an important statement as many haute couture creators have a keen understanding of the complex nature of women and capture that essence in their creations — especially in the timepieces that grace the wrists and track every moment. Indeed, as the worlds of fashion and timepieces coincide, haute joaillerie and horlogerie find new expressions of time that only fashion houses could envision, and the designer wristwatch becomes a sensational platform of time reinterpreted.

Gucci 18 karat white gold and diamond Chiodo watch

"Happiness... it lies in the joy of achievement, in the thrill of creative effort."

— Vincent Van Gogh

Diamonds are part of the Tambour legend and the Louis Vuitton spirit. They illuminate and highlight a dial, and in the Louis Vuitton Spin Time watch, they are an integral part of the poetry. The Tambour Spin Time utilizes a system of rotating disks to tell the time instead of hands — leaving the dial free for a story all its own — a moment for a woman to be captivated simply by the beauty of it all. At any given time, the rotating disks are all blank — save for the one that tells the hour. Invented by Louis Vuitton's masterful watchmakers and demonstrating the brand's exceptional know-how, the precious watch is set with black and white diamonds in the brand's iconic Monogram flower "stars" motif.

Hermes iconic H-Hour watch

"Time is a resource you should use to do something beautiful."

— Luc Perramond, CEO, Hermes

"I try to contrast; life today is full of contrast. We have to change."

— Gianni Versace

Versace Destiny Precious with yellow sapphires

H. Stern, DVF Sutra watch

"A watch should be designed like a gem and be the perfect everyday timepiece for the woman with a powerful and feminine spirit."

— Roberto Stern,
President and
Creative Director,
H. Stern

Nothing is Impossible.

"Freedom is what allows me to express my creativity in the products that I create. Technical mastery, expertise and know-how combined with a 'nothing is impossible' mantra created Crazy Carats."

— Sylvia Venturini Fendi, Creative Director, Fendi Accessories

Crazy Carats creates three dramatic watch looks in one timepiece for the modern woman. Change the stones at 11 of the 12 hour markers with a simple rotation of the crown. A diamond is maintained in a stable position at 12. The patented system features a total of 34 gemstones in each timepiece.

Haute couture and haute horlogerie come full circle with the Dior Christal Haute Couture Passage watches with elements of the timepiece embodying various elements found in Christian Dior Couture. Known for brilliant colors, Dior brought fashion to life with vibrant purple, fuchsia, orange, citrus yellow and acid green, and these alluring tones are emulated in the stunning Passage timepieces. To do this, the brand hand set each of the timepieces with hundreds of natural, untreated gemstones, such as this Passage N°3 watch set with bold green tsavorites. The oscillating weight of the Zenith Elite 671 mechanical movement is decorated with mother of pearl and gemstones and emulates the dial pattern, which is typically inspired by a Dior fabric. This Passage N°3 is crafted using 192 tsavorites weighing 15.92 carats and 103 diamonds weighing 0.52 carats.

"Elegance is an ensemble where the invisible is as important as the visible."

– Christian Dior

GEMS OF THE EARTH

The effects of color on time, as they pertain to the earth's most precious gems, are magnificent. Gemstones possess a power all their own and since the dawn of man have been used by cultures around the world for their spiritual qualities and unending beauty. Wars have been fought over gems, hearts won over because of them and powers bestowed unto rulers with them. It is only fitting, then, that they adorn the finest of women's watches. From regal sapphires to rubies, to the vibrant colors of peridot, topaz, citrine and amethyst, gemstones take center stage in today's timepieces, offering flawless, artistic creations that color time with an unrivaled painter's palette.

Piccolino watches by de Grisogono are created in a breadth of nature's finest gemstones.

This Infinity Curvex watch from Franck Muller is meticulously set with more than eight carats of diamonds and gemstones on the case and dial. It is part of the brand's exotic Four Seasons collection.

Q&A with Franck Muller, Co-Founder

What do you think of when you design a watch for a woman?
The words that come to mind when designing for women are nature, femininity and elegance.

How do you think women use or see time?
Beauty is a fleeting concept that women in particular are well aware of. Therefore, they attach great importance to time and organize their lives in order to enjoy every minute of it.

What do you think a woman wants in a watch?
Previously it was mostly men who attached great importance to watches and complications. However, times change and so do attitudes. Women today are more interested in watches. They seek a timepiece aesthetically suited to their tastes and needs, but they also want a timepiece worthy of its name; something that reveals the beautiful savoir-faire of the watch making.

"A woman has four seasons in her life: a period of discovery of the world, a period of self-assertion, a period of recognition and a final stage of internal peace. Their choices reflect their tastes and feelings between these transitions."

— Franck Muller, Co-Founder and Master of Complications

This elegant Four Seasons collection Franck Muller Double Mystery watch features two rotating discs to display the hours and minutes via an automatic movement with segmented platinum rotor. The watch bezel is set with 259 rubies, emeralds, green tsavorites, spessartite, amethyst and sapphires in blue, yellow, orange, pink and violet weighing 3.6 carats. The dial and case are set with 338 diamonds weighing 2.59 carats and 14 colored gems accent the dial.

Ritmo Mundo Persepolis is an innovative round watch that revolves in its case. The two-sided timepiece is created in a fantastic array of color-gemstone-set bezels that are sure to delight anyone.

" A watch gives you the time, but it's up to you to stay on time, to achieve what you want."

— Ali Soltani, CEO, Ritmo Mundo

This Bertolucci STRIA III watch is crafted in stainless steel and 925 sterling silver and is set with 247 diamonds and 77 blue sapphires.

"The value in life lies not in the length of days but in the use we make of them..."

— Michel de Montaigne

Corum's Classical Billionaire Tourbillon is crafted in 18-karat white gold and features a magnificent mystery skeletonized movement with tourbillon escapement to compensate for errors in timekeeping due to the effects of gravity on the watch in different positions. The masterpiece features 994 full-cut diamonds weighing 46.86 carats and 162 blue sapphires weighing 22.78 carats.

From Jaeger-LeCoultre, the Reverso Squadra Art watch is ablaze with diamonds and colored sapphires for a radiant appeal. The snow setting of the gems in graded colors offers alluring beauty on the front and back of the watch. This extraordinary setting is matched by an equally exceptional mechanism, the mechanical Jaeger-LeCoultre Calibre 822.

"A thing of beauty is a joy forever."

— John Keats

Graham's Swordfish Ali Baba has a limited production of 40 exclusive pieces set in colorful precious stones. Following the iconic Swordfish in stainless steel, the Ali Baba features variously-sized emeralds, rubies, sapphires and diamonds exquisitely set on the watch with a total weight of 4.65 carats. A dash of 40 mandarin garnets complements the precious rainbow of gemstones.

"Think for a moment about how rare it is to have time to do the things you enjoy most, and you will realize how special 'time' can be."

— Richard Mille, President & CEO, Richard Mille

This Richard Mille RM 007 Ladies Automatic is crafted in 18-karat gold, with 104 rubies and 142 pink sapphires, totaling 1.35 carats.

Q&A with Ivanka Trump | Jewelry Designer

Do you think women see time differently than men?

I think women tend to map out their days more rigidly than men do. As a woman, I am constantly put into situations where I have to multi-task and focus on multiple work related projects while also taking care of my home and the needs of my family. I think men tend to be less schedule-bound.

What do women have over men when it comes to appreciating time and watches?

I believe women appreciate every minute of every day. I stick to a very strict schedule to make sure that my time is utilized to its fullest so I am able to balance my career and my family. I am constantly looking at my watch to keep on track... or at least attempting to.

Do you think women appreciate watches more so than men?

I think that men and women appreciate different features of watches... I personally love timeless elegant watches that have sleek clean designs. I invest in timepieces that I know will one day be worn by my daughter. Watches are like time in the sense that men and women find certain things more significant than others.

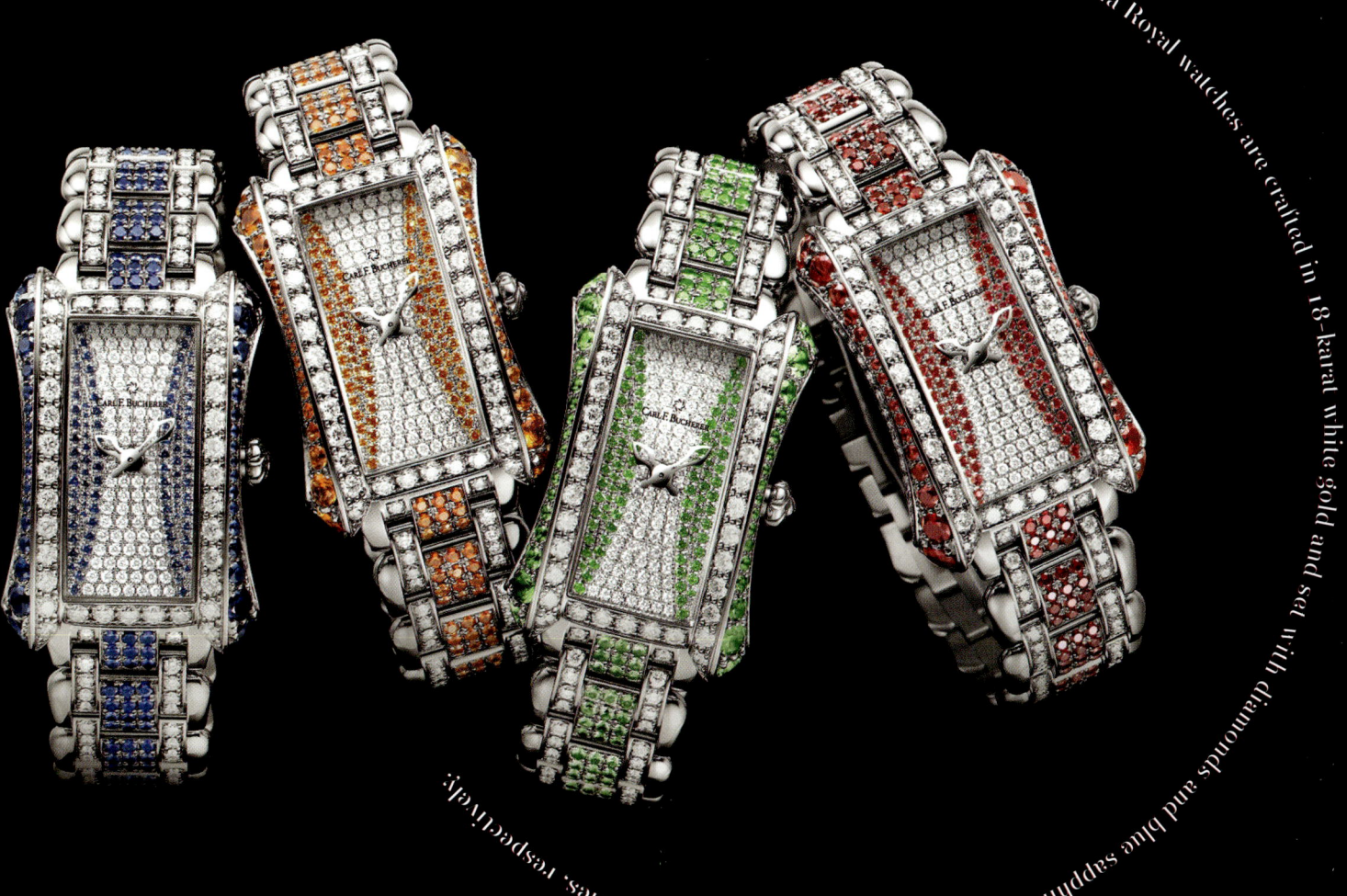

Carl F. Bucherer Alacria Royal watches are crafted in 18-karat white gold and set with diamonds and blue sapphires, orange sapphires, emeralds and rubies, respectively.

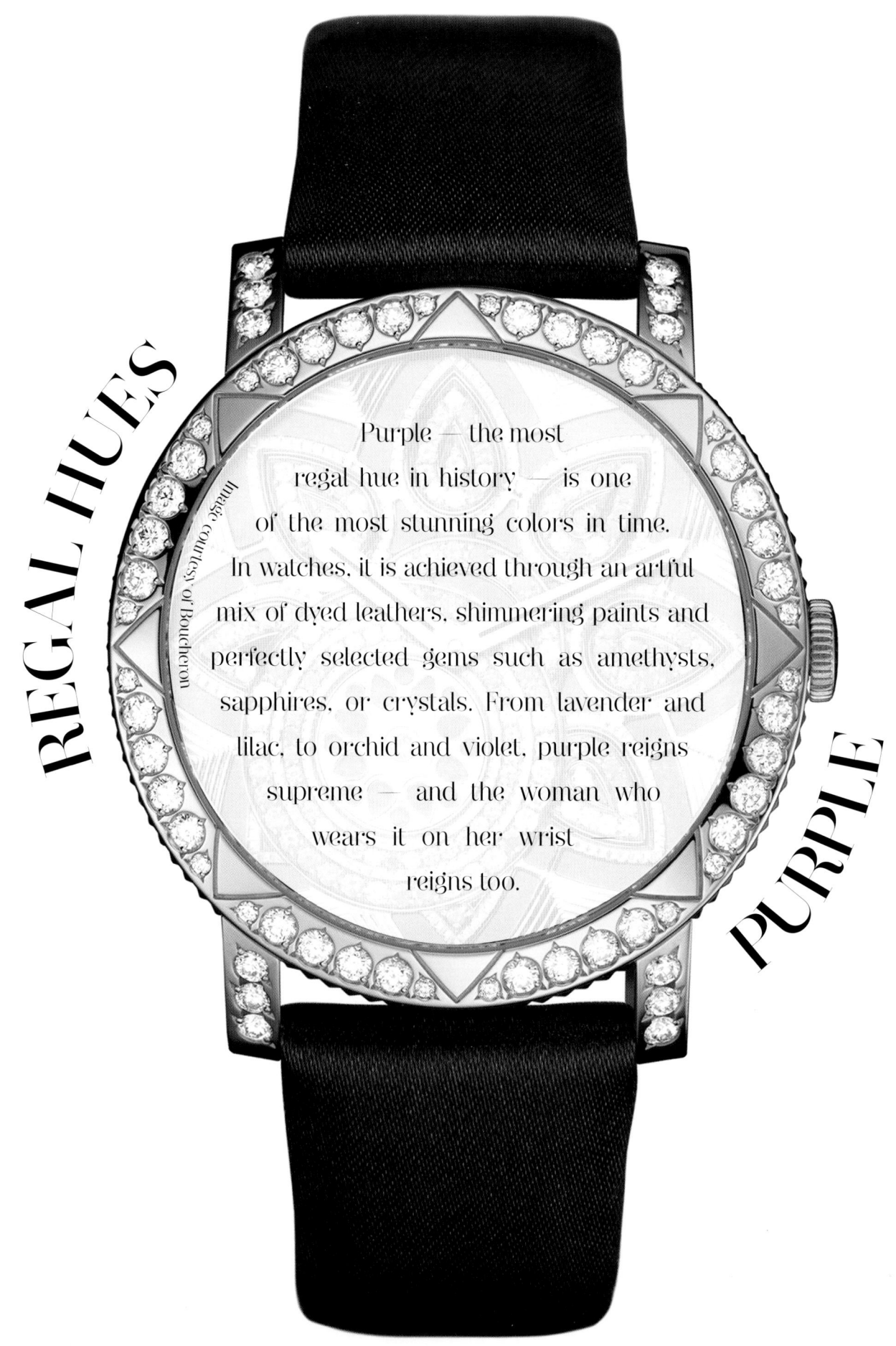

REGAL HUES PURPLE

Purple — the most regal hue in history — is one of the most stunning colors in time. In watches, it is achieved through an artful mix of dyed leathers, shimmering paints and perfectly selected gems such as amethysts, sapphires, or crystals. From lavender and lilac, to orchid and violet, purple reigns supreme — and the woman who wears it on her wrist — reigns too.

Image courtesy of Boucheron

"Womanist is to feminist as purple is to lavender."

— Alice Walker

This Concord C1 Chronograph Amethyst is a masterpiece both inside and out. The haute Joaillerie watch is set with 217 baguette-cut amethysts totaling 13.80 carats — and offers distinctive lines, rounded curves and a singular splendid look that does not give away the fact that the watch houses an automatic COSC-certified Chronometer chronograph Valgrange caliber that beats at 28,800 vibrations per hour. The 25-jewel movement features a snailed rotor and Côtes tde Genève motif. It offers 48 hours of power reserve and chronograph function. The watch case is crafted in 18-karat white gold and is water resistant to 30 meters. The strap is purple alligator and even the clasp and the crown are set with baguette-cut amethysts.

Hublot Big Bang Amethyst

Q&A
with Jean-Claude Biver, CEO, Hublot

What do you think the relationship is for women and time?

For a woman, time should not exist, and if it exists, it should be eternal.

Is a watch the soul of time that we capture on our wrist?

Every watch made by hand and with love has a soul. A soul is given by the watchmaker through his passion and the love of his work. Only a few brands and watchmakers are capable of giving birth to a soul, as well as only a few people are able to understand and capture the soul of a watch.

How is a watch like a woman?

The parallel we have between a watch and a woman is first the beauty, but more importantly, the hidden beauty. The inner beauty is what makes a woman such a unique Lady, and which makes a watch such a unique piece of art.

"Strong design and feminine details in watches embody the many facets of today's woman. Dynamic and expressive, yet soft and thoughtful."

— Lionel Favre, Head of Design, Roger Dubuis

The Roger Dubuis Excalibur Lady Amethyst houses a self-winding, 33-jewel mechanical movement made in-house in Geneva that is a COSC-certified chronometer. Its lilac mother-of-pearl dial is made of natural mother-of-pearl that is darkened with a purple varnish on the back of the pearl via a plate. Approximately 10 steps are required to realize such a dial, as the raw material remains untouched. The bezel is set with 48 baguette amethysts. There are just 88 pieces made of this watch.

This stunning beauty is an owlish delight outside and a magnificent work inside, as well. Conceived of in collaboration with the legendary Boucheron (who has also co-partnered in timepiece designs with Girard Perregaux), the MB&F JWLRYMachine is predicated on the HM3 platform – housing the superb movement that made its debut first as a magnificent new timepiece, then interpreted by innovative watchmaker Max Busser as a frog and now as an avant-garde bejeweled beauty. This solid gold rendition is set with amethyst and violet sapphires with diamonds.

DAZZLING DIAMONDS

A veritable winter wonderland, diamond wristwatches are in a league of their own. Created in everything from simple diamond timepieces to lavish million-dollar extravaganzas, diamond jewelry watches are works of art and craftsmanship that turn time into glamour. Often these stunning eye catchers can take months to create, with stonecutters and gemsetters working together to find the best match in color, cut and clarity. A host of settings are used in diamond watches — running the gamut from channel set (where stones are set one next to another with no metal showing) to invisibly set (where the watch looks as though it is entirely created of diamond after diamond with absolutely no gaps), prong-set (typically done with diamond brilliants) and pavé set (with many miniscule diamonds put together for a paved effect). Today, many astute watchmakers have even developed their own artisanal diamond-setting techniques to remain ever more unique. Depending on the complexity of the setting of the diamonds and the design of the watch, it can take hundreds of hours to complete a piece. One famous watch took more than 6,000 hours to select and set the diamonds. Due to the exceptional work that goes into the making of these diamond watches, most are heirloom pieces — meant to be handed down from one generation to the next — making time as timeless as we all wish it could be.

The popularity of diamond watches is never going to pass, especially with extraordinary timepieces such as these masterpieces built by Patek Philippe, which have transformed the ultimate diamond bracelet into revolutionary jeweled trackers of time.

"Guard well your spare moments. Discard them and their value will never be known. Improve them and they will become the brightest gems in a useful life."

— Ralph Waldo Emerson

Audemars Piguet Millenary Précieuse Diamond manual winding wristwatch with more than 300 baguette-cut and brilliant-cut diamonds weighing more than five carats.

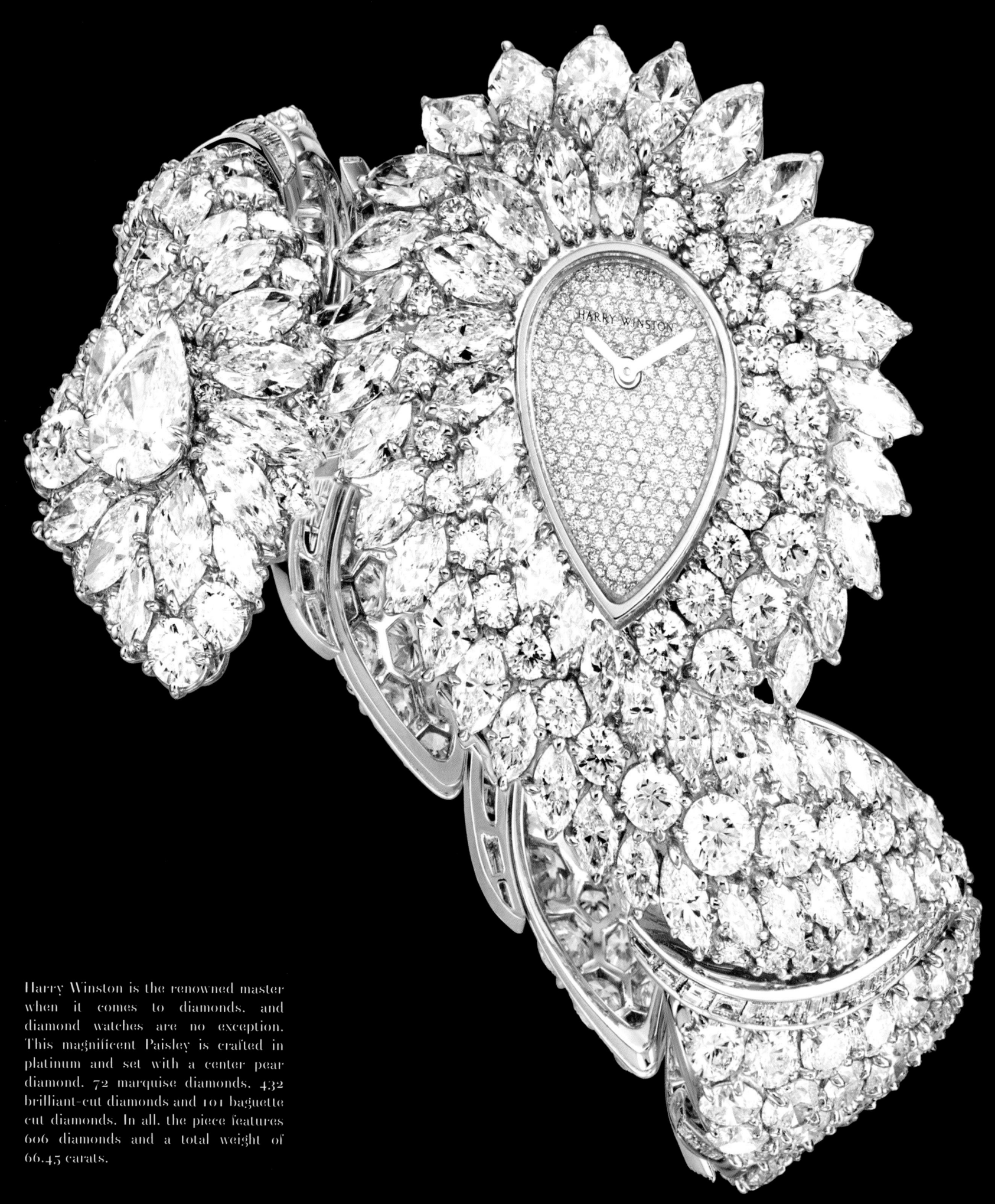

Harry Winston is the renowned master when it comes to diamonds, and diamond watches are no exception. This magnificent Paisley is crafted in platinum and set with a center pear diamond, 72 marquise diamonds, 432 brilliant-cut diamonds and 101 baguette cut diamonds. In all, the piece features 606 diamonds and a total weight of 66.45 carats.

Men grow cold as girls grow old

And we all lose our charms in the end

But square-cut or pear-shaped

These rocks don't lose their shape

Diamonds are a girl's best friend

Tiffany's

Cartier

Talk to me, Harry, Winston

Tell me all about it!"

— Marilyn Monroe in

Gentlemen Prefer Blondes

"Perhaps time's definition of coal is the diamond."

— Kahlil Gibran

Damiani Mimosa in 18-karat white gold set with 159 diamonds weighing 4.43 carats

Vacheron Constantin Lady Kalla Flame dances to life by the brilliance of its flame-cut diamonds. The 18-karat white gold case emulates the flame shape of the gems. The bracelet is a natural extension of the case with its sensual design that distracts one from the fact that the watch houses a mechanical manual-wind movement.

The case of the Vacheron Constantin Lady Kalla Flame is set with 20 flame cut diamonds weighing approximately 10 carats. the dial is set with 60 flame cut diamonds weighing two carats and the bracelet is set with 120 flame cut diamonds weighing 24.50 carats.

The entire masterpiece is carved by hand in order to fit the gemstones into the gold frame and then the 36.5 carats of gems are placed individually by hand over the course of hundreds of hours.

This Piaget Limelight Exceptional Piece. Haute Couture Inspiration watch is created in 18-karat white gold with 1,210 brilliant and baguette-cut diamonds and black spinels.

"The required creativity in jewelry watches mirrors Piaget's belief that women consider watches as a sublimated piece of jewelry and a true expression of their personality."

— Philippe Leopold-Metzger,
   CEO, Piaget

SHAPELY SENSATIONS

Thank goodness watchmakers were never content with the simple round watch. Round evolved into square, rectangular, tonneau (barrel shaped) and eventually into a myriad of wonderful shapes that reflect the woman's feminine nature. Everything from cylindrical lipstick like cases to curved figure eights, hearts and asymmetrical delights take form in today's timely wonders. But make no mistake — these shapely sensations are no easy feat. Creating curvaceous cases and appropriately adorning them with diamonds and gemstones to make them ever more appealing is no easy task. Still, the most innovative brands rise to the challenge and do so with shapes that keep us spellbound. Round is still luscious, but every woman knows that truly good things come in all sizes and shapes.

For more than a century Van Cleef & Arpels has been creating some of the most instantly recognizable timepieces in the world. With the famed Cadenas watch, haute joaillerie comes together with the watchmaker to make magic in time. This famed series was first unveiled in 1935, inspired by the Duchess of Windsor, and has been an important, emblematic classic ever since. This all-diamond, 18 karat white gold Cadenas Haute Joaillerie watch recalls the splendor. This Cadenas all-diamond watch is set with 749 Round white diamonds weighing 19.5 carats and 122 Baguette-cut white diamonds weighing 13.86 carats.

This Blancpain Saint-Valentin 2010 watch is crafted in 18-karat white gold and features a pink and white mother-of-pearl dial. It is set with nine round diamonds and a heart-shaped diamond on the dial. The bezel is set with 404 diamonds and 66 pink sapphires; the crown is set with a cabochon-cut pink sapphire. The watch houses a self-winding movement.

"In art the hand can never execute anything higher than the heart can inspire."

— Ralph Waldo Emerson

This elegant Ralph Lauren Diamond Link Stirrup Watch is meticulously crafted in 18-karat white gold and features a full pavé diamond case and chain-link bracelet set with 1,553 diamonds for a total weight of 24.7 carats. The manual-winding mechanical movement caliber RL430 is made by Piaget for Ralph Lauren.

Q&A

with Fawaz Gruosi, Founder, Designer, de Grisogono

When you design for a woman, what do you think of?

To me, originality is key. It is the most valuable thing we can possess when we are designing. Being different must always come first.

You are a leader in timepieces with black diamonds. Why this gem for watches?

People are often suspicious of unfamiliar things, a trap I almost fell into myself in the beginning. Once I started creating in black diamonds, I have never stopped, and I never will. There is something too special about the black diamond and the place it holds in time.

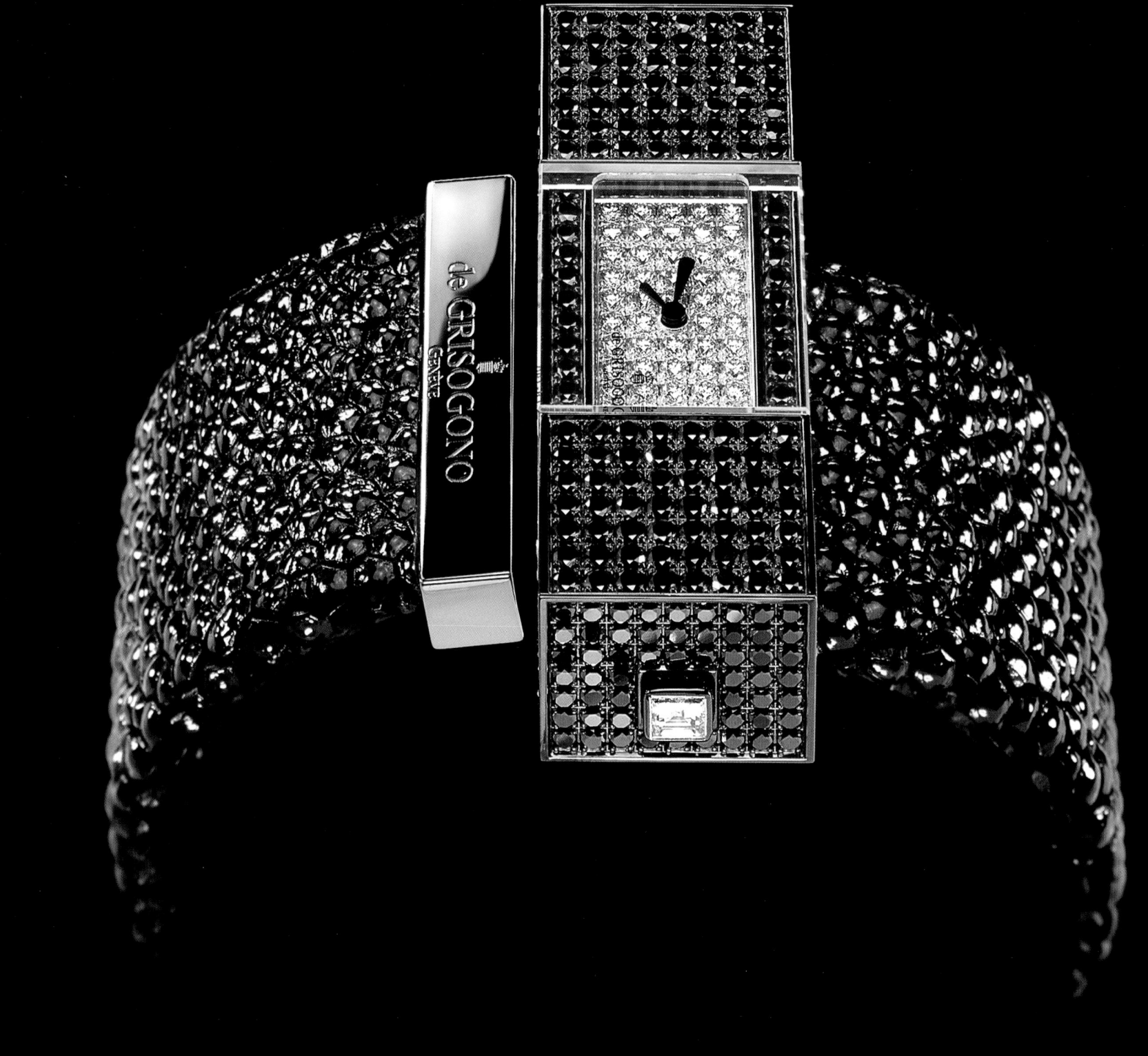

Irresistibly feminine, the de Grisogono Lipstick jeweled watch is based on a cylindrical concept that is squared and harmoniously balanced. This version is crafted in 18-karat blackened gold and meticulously set with 503 black diamonds weighing 10.03 carats. The dial is polished white gold and set with 54

Breguet, the brand founded by famed Abraham-Louis Breguet, creator of watches for Marie Antoinette, continues to surprise and delight watch connoisseurs with its masterful skill not only of watch movements but also of watch sculpting and design. This Les Jardin du Petit Trianon-Les Glycines watch is crafted in 18-karat white gold and features a bezel, dial flange, case and lugs paved with 284 diamonds totaling 3.81 carats. The stunning bow motif is set with six pear-shaped diamonds weighing 1.42 carats and paved with 71 diamonds representing another 4.17 carats. The movement is a self-winding caliber.

> "It's kind of fun to do the impossible."
>
> — Walt Disney

FLORAL MASTERPIECES
TIME IN BLOOM

From scent to color to touch and look, few things in life excite so many senses as flowers. These exotic wonders therefore translate elegantly in all forms onto watch cases and dials. Whether interpreted in paintings or in gemstones, with diamonds or with color, floral watches are among the world's most scintillating beauties. Some brands even extend their floral canvas from the dial to reach around the case, as well, or even out on to the bracelet with forms that replicate vines and flowers of the most entrancing kind. These pieces are truly heirloom watches — art at its finest meant to be passed down for generations to come.

The Poppy Flower Timepiece by Andreas von Zadora-Gerlof

Q&A

with Andreas von der Goltz

How is a watch like a woman?

Throughout the ages women have been revered in art. Much like the artist transforms a piece of marble into a beautiful female form, we take precious and semi-precious stones and assemble them into a beautiful work of art which in turn graces the wrist of its wearer.

Your watches are more like sculptures on one's wrist—what kind of woman buys a Zadora timepiece?

A woman who buys a Zadora Timepiece is adding a major piece of jewelry to her collection. Our watches are possessions to be kept forever and passed on to the next generation. One cannot put a price on the value of an artistic piece and something primarily appreciated by the eye of the beholder. All of our watches are one of a kind, unique custom designs taking 2-3 months to create by hand, making them highly desirable.

How is it different designing for women exclusively, compared to men's watches today?

Because our designs focus mainly on women, we are able to create magical and creative pieces of jewelry to be worn mostly on special occasions. Unlike men, women have a great variety of jewelry to choose from whereas men primarily only have one, that being a watch. Therefore, we take great interest and pride in having our timepieces be different from that of any other woman's watch. With their unique design we are indeed "changing the face of time" by taking the traditional watch face and turning it into a platform for an exquisite work of art.

What do you think women want in a watch?

We believe women want a sensational objet d'art that has the added bonus of telling time.

The Zadora Poppy Flower Timepiece is executed in platinum and palladium, 18 karat white gold. the flower is meticulously pavé set with pink and white diamonds, rubies, and pink sapphires, giving an ombré effect to its voluptuous petals with its pistils minutely textured in black patina on platinum.

Emeralds and diamonds bespeak fanciful elegance in this Piaget Limelight Garden Party watch with Leaf Inspiration. It is magnificently created in 18 karat white gold with 34 brilliant cut diamonds and 10 marquise cut emeralds forming the leaves. The dial is set with 76 brilliant cut diamonds. It houses a Piaget 56P quartz movement.

"Flowers are the sweetest things God ever made and forgot to put a soul into."

— Henry Beecher, Life thoughts, 1858

Inspired by Lewis Carroll's poem in *Alice's Adventures in Wonderland – All in the Golden Afternoon –* DeWitt develops a poetic and artistic rendition of a woman's perceptions of the world. In the poem of Alice's Adventures in Wonderland, Lewis Carroll recalls the afternoon that he took a boat trip from Oxford to Godstow and improvised the *Alice in Wonderland* story to entertain the three Liddell sisters, Lorina, Alice and Edith.

The DeWitt Golden Afternoon collection features three models that reflect the journey from innocent childhood to wisdom and adulthood as seen by the creators at DeWitt. DeWitt's interpretation is a dial of delicate mother-of-pearl with colorful flowers across it in different poetic tones. The slightly "out of focus" effect reminds one of a photographic style, powerful and evocative. The hours and minutes hands are refined little sculptures of angel wings, while the seconds hand is topped with a flame.

"All in the golden afternoon

Full leisurely we glide;

For both our oars, with little skill,

By little arms are plied,

While little hands make vain pretence

Our wanderings to guide."

— Lewis Carroll

"A humble flower is the work of centuries."

— William Blake

Bovet, established in 1822, is under the direction of Dimier 1738 Manufacture, an artisan manufacture of Haute Horlogerie situated in Tramelan, Switzerland. Here the company creates exceptional timepieces in house, from research to development, offering state-of-the-art handcrafted movements and exquisite dials. The Bovet Fleurier Amadeo "Mille Fleurs" watch features a case concept that is the result of seven years of research, and enables the wearer to transform the timepiece from wristwatch to table clock to pocket watch or pendant watch extremely easily. The dials are all hand-painted on mother-of-pearl.

Fourth generation Piaget family member Yves Piaget is not only a superb diplomat and staunch Piaget jewelry and watch forger into the future, but also a true flower connoisseur; leading the foray into the creation of Piaget floral watches as early as the 1960's. It is a tradition the brand carries on today, creating some of the most superb floral inspired timepieces.

It is interesting to know that even real flowers live on in the Piaget name, as Yves Piaget has a famous rose named in his honor. The Meilland family introduced one of the worlds most romantic and glamorous hybrid tea roses in 1985, the Yves Piaget Rose — a delightful deep pink beauty with a sweet, fruity fragrance. Bloom on, Piaget.

Piaget Miss Protocole XL watches in 18-carat white gold with diamond pave and embellished with a floral motif in 'grand feu' champleve enamel on an interchangeable strap.

"Being
perfect artists and
ingenuous poets, the Chinese have
piously preserved the love and holy cult of flowers:
one of the very rare and most ancient traditions which has
survived their decadence. And since flowers had to be distinguished from each other, they have attributed graceful analogies to them, dreamy images, pure and passionate names which perpetuate and harmonize in our minds the sensations of gentle charm and violent intoxication with which they inspire us. So it is that certain peonies, their favorite flower, are saluted by the Chinese, according to their form or color, by these delicious names, each an entire poem and an entire novel: The Young Girl Who Offers Her Breasts, or: The Water That Sleeps Beneath the Moon, or: The Sunlight in the Forest, or: The First Desire of the Reclining Virgin, or My Gown Is No Longer All White Because in Tearing It the Son of Heaven Left a Little Rosy Stain; or, even better, this one: I Possessed My Lover
in the Garden."

— Octave Mirbeau, *Torture Garden*,
"The Garden," Chapter 5

The Bedat & Co. No. 8 floral watch is crafted in solid palladium and white gold and features an open-worked case. It is masterfully set with 243 diamond brilliants. The satin strap features a buckle set with 66 diamonds for a total weight of 3.41 carats.

This Chanel "Camelia Ruban" watch is crafted in 18-karat gold and meticulously set with black and white diamonds.

"Flowers whisper 'Beauty!' to the world." — Dr. Sun Wolf

From Jaeger-LeCoultre these stunning La Rose and Yellow Tulip watches are secret watches that unfurl their petals to reveal the watch dial beneath. The rose is created with 75 carats of gemstones, including 1,480 pink sapphires, 1,370 tsavorites, 120 rubies and 120 diamonds. The tulip consists of 85 carats of gems: 1,600 tsavorites, 1,900 yellow sapphires and 30 diamonds. Each watch houses the JLC Caliber 846 manual-wind movement.

CALL OF THE WILD

Beloved animals have long influenced the world of art, and have entranced the hearts and souls of women and men alike. Thus it comes as no surprise that watch companies should choose these friends (and foes) as inspiration for decoration and adornment. From sea animals to domesticated pets, from desert mammals to jungle dwellers and the wildest cats of the world — animals evoke an unparalleled emotion. Enchanting, alluring and mysterious — animals escape the essence of time becoming ethereal witnesses to nature — past, present and future. Today's watchmakers make every effort to capture these beasts and forever immortalize them on the wrists in some of the most traditional forms of art. Recreating these animals in enamel, marquetry, mosaic or magnificent gemstone splendor is an amazing challenge when one considers that these artisans are embodying tiny details, as well as spirit and emotion, on a dial (or canvas) that is just slightly bigger than an inch in diameter.

This Hublot Big Bang Leopard is an enticing mix of sexy, ultra feminine glamour thanks to the mechanics and design. It features a leopard print denim strap stitched onto black rubber for a softer feel and chic appeal. The dial is a realistic leopard print and is accented with eight yellow diamonds. The 41mm 5N rose gold case features a satin finished rose gold bezel set with 48 andalusites, smoked quartz and baguette citrine gemstones that brings the entire leopard print colors together seamlessly. It houses the HUB 4300 automatic mechanical chronograph movement.

"The Secret of Life is in Art."

— Oscar Wilde

This entrancing Cartier Tortue XL watch in yellow gold features a jaguar motif in relief engraving and champlevé grand feu enamel that recalls the great tradition of artistic craft. Almost leaping off the case, the jaguar is strong and bold with detail and highlights, depth and dimension from the deep-set eyes to the jaw and muzzle. Mesmerizing is the operative word in this timepiece. In fact, one may forget all about wanting to check the time. However, Cartier hasn't forgotten, and has equipped the watch with a manufacture mechanical manual wind caliber 9601 movement. The watch requires more than 60 hours for the engraving and assembling, followed by another 25 hours for the champlevé grand feu enameling with its seven shades that pay homage to the majestic cat. Just 80 individually numbered pieces will ever be made. (© Ines Dieleman / Cartier)

Few can resist the allure of the call of the beastly tigers and lions that haunt our dreams and hunt our wild side. Certainly Ulysse Nardin — makers of so many wonderful Safari Jaquemarts watches — could not turn itself away from the majesty that is nature. This emerald-eyed tiger beckons to the wearer and all about her; enticing one to come just a little bit closer. As though alive, she crouches in waiting, aglow in diamonds and enamel and surrounded by ruby markers. In Ulysse Nardin fashion, the 18-karat white gold Caprice Tiger watch houses a UN mechanical self-winding movement.

The Ulysse Nardin Caprice Tiger watch is crafted in 18 karat white gold and features a tiger motif dial of diamonds and emerald eyes.

"Diamonds and time are your best friends."

— Caroline Gruosi-Scheufele, Co-President and Artistic Designer, Chopard

In true Chopard style, the brand regularly rises to the challenge of creating the audacious and alluring. In its Animal World collection, Caroline Gruosi-Scheufele, co-president of Chopard, was first inspired by plush animals, soft teddy bears and stuffed toys that one carried with them as children. She reinterpreted these comfort animals into wearable jewelry and watch works of art to always be cherished and to remind us of nature that surrounds us. To create these special editions, the company draws on the full range of in-house talents, mingling tradition with cutting-edge techniques and meticulous research to express excellence and beauty.

Cartier has long been known for its extraordinary works of high jewelry and haute horology. A pioneer for more than 160 years in the watch and jewelry world, Cartier has always focused on artistry in its designs. In the 21st century, the brand has created masterful new calibers in high watchmaking that has brought it to the forefront of complex watchmaking. However, one cannot forget that this is a brand rich with artistic prowess and creative excellence, and in the past decade has intensified that spirit in its Cartier Art collection of unique watches. Cartier has unveiled a series of timepieces that clearly demonstrate the brand's unparalleled mastery of the traditional arts: marquetry, mosaics, engraving and enamel. Each watch is more stunning and alluring than the next. Such is the case with this gem set and Grand feu champlevé enamel Ballon Bleu de Cartier watch in 18-karat white gold with monkey motif.

The Champlevé enamel is applied to the cells that have been engraved by the Cartier craftsmen according to the design. Multiple paintings and firings (at about 800 degrees C) are required to create a single dial, which teaks approximately 36 hours of highly precise work. The painting is then further emblazoned with a coat of 225 cognac diamonds for the monkey, and an outer bezel of 126 round white diamonds. The watch houses a self-winding mechanical Caliber 049. Just 50 of this mini miracle will ever be made, each one unique due to the hand painting. (© Ines Dieleman / Cartier)

(Left:) Enameling the dial in an assortment of greens. (© Ines Dieleman / Cartier)

(Right:) Enamel powders (© Ines Dieleman / Cartier)

Cartier is a master at haute horology and haute joaillerie — with certain masterpieces being a blend of both of these worlds and more. Such is the case with the Cartier Art series of unique and individualistic pieces. This magnificent Rotonde de Cartier 42mm Tortoise watch, crafted in 18-karat pink gold and set with 1.39 carats of diamonds on the bezel, houses a Cartier Manufacture manual wind movement. The true work of art and craftsmanship comes on the dial, however — with the turtle motif entirely created by hand — a mosaic of more than 1,000 stones. To build this masterpiece, a stone setter works with tweezers and ultimate patience — masterfully setting each and every mosaic stone into place, one at a time, arranging the colors with an eye for every single detail — to finish the dial in such a way that makes the tortoise one of the most realistic and alluring mosaic dials ever.

Just 10 individually numbered pieces of the Rotonde de Cartier 42mm Tortoise watch will ever be created. (© Ines Dieleman / Cartier)

(Above:) In 2011, Cartier unveiled a series of six timepieces that clearly demonstrate the brand's unparalleled mastery of the traditional arts: marquetry, mosaics, engraving and enamel. (© Philippe Gontier / Cartier)

(Left:) The stones in this superb Rotonde de Cartier 42mm Tortoise mosaic watch are onyx, tiger's eye, falcon's eye, yellow pietersite, carnelian, three colors of jasper, yellow and moss agate, coral and mother of pearl. No less than 60 hours of work are required to position the stones on this infinitely small scale; then they are cemented to ensure an even surface and an extraordinary fantasy comes to life in a mineral turtle that defines time. (© Philippe Gontier / Cartier)

A man who works with his hands is a laborer; a man who works with his hands and his brain is a craftsman; but a man who works with his hands and his brain and his heart is an Artist."

— Louis Nixer

The 1,167 squares that make up the mosaic are positioned one by one.
(© Philippe Gontier / Cartier)

From Boucheron's Exotic Parade series, this stunning Crazy Hathi watch honors the Indian elephant, which is ceremoniously dancing across the dial, trunk proudly curled up in the air. The watch recalls Louis Boucheron's first trip to India in 1909, when he became enamored with the land and its culture. Since that time, a style emerged from the House that has become inextricably tied with India and its culture. This alluring Crazy Hathi is crafted in 18-karat pink gold and set with yellow and orange sapphires.

Elephants have long held a special meaning in the worlds of culture and religion. For some, they are a sacred animal; for others they are the seer of all. Even today, elephants have a place of prominence in certain religions. They are the animals chosen to carry relics of grace in the sacred processions in Buddhism. These great and wondrous creatures are often the center of myth and fascination, of enchantment and charm. Different species exist, and are found in different parts of the world — and each is revered for its own special reason. The likeness of these animals makes great good luck charms, and the emulation of them on timepieces by some of the finest watchmakers in the world, such as Boucheron, is inevitable, as they are to be preserved, protected and treasured.

Van Cleef & Arpels regularly seems to stun the world with its unveiling of incredible objets d'art in the high jewelry collection of timepieces like its most recent creation inspired by Jules Verne's books of journeys, ballooning and magical travel. The watches mime a journey in a balloon around the world, and thus certain dials depict animals and sights one would see in different continents. Each is a work of art more beautiful and compelling than the next. It is timepieces like these that have kept this French Maison always at the forefront of hearts and minds of watch lovers. Indeed, not only is it the allure of the timepieces, and the extreme exclusivity, but also it is the exceptional craftsmanship that goes into their making. These magnificent sculpted watches trace the traveler's discovery of the world from the balloon's basket, where he looks over the animals and vegetation of the lands.

The Van Cleef & Arpels Lady Arpels unique styles employ amazing artistic talent combining the sculpting of the gold that captures the tenderness of the animals, and the placement of the mother-of-pearl inlays, champlevé enamel and diamonds. It takes Van Cleef & Arpels' master artisans hundreds of hours to perfectly carve the settings for the diamonds, to engrave the gold with the features and nuances of each animal to the most minute detail, and to finish and polish each piece perfectly. Just 22 pieces of each of these wonders will ever be made—each one unique by the very fact that each is hand sculpted.

Lady Arpels African landscape giraffe décor cased in 18-karat white gold, set with diamonds. Dial in mother-of-pearl inlays, sculpted white gold and champlevé enamel set with diamonds. Limited edition of 22 pieces.

This Lady Arpels Polar Landscape White Bear is crafted in 18-karat white gold, and set with diamonds. The dial is in mother-of-pearl inlays, sculpted white gold and champlevé enamel set with diamonds. It is created in a limited edition of 22 pieces.

© BNvision SA/Joseph BATO (Courtesy of Van Cleef & Arpels)

The Chopard Animal World Penguin watch features a penguin of black and white diamonds on a mother of pearl marquetry background.

> "A picture is a poem without words."
>
> — Horace

"Without art, the crudeness of reality would make the world unbearable."

— George Bernard Shaw

The Chopard polar bear Animal World features a three-dimensional polar bear on a backdrop of white mother-of-pearl emulating the snow.

"Take time to deliberate; but when the time for action arrives, stop thinking and go in."

— Napoleon Bonaparte

This Zadora Snake Timepiece is executed in patinated palladium 18-karat white gold and pavé set with black and canary yellow diamonds and pear shaped emerald eyes, the serpent clutches a baroque South Sea pearl in its mouth.

SECRETS IN TIME

Few watches captivate like the secret, or hidden, watch. Truly a piece of jewelry, these timepieces entitle women to their own private rendezvous with time. Initially created to protect watch dials from damage, the concept of the covered watch fast became a design theme — an ornamental interpretation of life and style. Today, many watch brands use the covered watch as a means to offer versatility and fantasy — to bring magic and mystery to the forefront. No longer are these watches just timekeepers, they are jewels, inspirational works of art and mystical, innovative interpretations of time. They also let a woman do what no other watch can — hide the passage of time by simply closing or reversing the cover. This luxurious genre of watches, seductive possessions, embodies secrets revealed, secrets hidden in time. Often secret watches are the most difficult timepieces to execute due to the complexity of their hidden dials. Hinged or reversing lids must work beautifully, without a snag. The mechanics of the timepiece must equal the elegance of the watch, and the craftsmanship must deftly hide the secret that lies beneath... time passing unnoticed.

Cartier is and always has been a master at creating hidden watches. This Secret watch with feather motif is from the Mille Et Une Heures Cartier collection. (© Cartier)

Since its founding in 1874, Piaget has been creating dazzling haute horlogerie timepieces that have won the hearts of royalty, celebrities and collectors around the world time and again. This is a brand whose reputation is built on daring difference, technological advancement, beguiling masterpieces and outstanding craftsmanship. Nothing is as daring and alluring as its secret watches, built to bewitch with enthralling beauty. This Exceptional Secret Ribbon Inspiration watch is crafted in 18-karat white gold with a total of 552 brilliant and baguette-cut diamonds. Entirely crafted by hand, the piece required 870 hours of development and 200 hours of gem setting.

"Secrets are made to be found out with time."

— Charles Sanford

The Harry Winston Rosebud Secret watch Harry Winston is a versatile and alluring piece: a diamond pendant watch that can be worn as a wristwatch, or as a brooch watch. It consists of 69 carats of brilliant, pear-shaped and baguette-cut diamonds, with a sliding cover to keep time a mystery.

"No two diamonds are alike. Each diamond has a different problem. Each diamond has a different nature. Each diamond must be handled as you handle a person."

— Harry Winston

"We dance round in a ring and suppose. While the secret sits in the middle and knows."

— Robert Frost

Bulgari Mediterranean Garden of Eden watch

"Secret watches are exceptional. Secrets are a tradition of high jewelry. For a long time it wasn't acceptable for women to wear watches, so when jewelers started to create women's watches, they hid them, as secrets so it wasn't apparent that a woman was looking at the time. When we create our secret watches, we hide the time in little drawers within the watches, like a secret drawer a woman can hide her treasures in."

— Nicolas Bos, President and CEO, the Americas,
Worldwide Creative Director,
Van Cleef & Arpels

Van Cleef & Arpels, Secret Duo Caresse d'Eole Fairy watch

The Zadora Scorpion Timepiece is executed in patinated 18-karat white gold and micro pavé set with grey and white diamonds, with ruby eyes and rubellite stinger; the scorpion crouches on a guilloche top.

"All women are natural born espionage agents."

— Eddy Cantor

"You have to kiss a few good frogs to find a Prince." —Graffito

The Zadora Frog Prince Timepiece is executed in textured 18-karat yellow gold, the body is pavé set with tsavorites studded with blue sapphire cabochons with diamond feet details and ruby cabochon eyes. This royal amphibian proudly wears a diamond-encrusted crown.

THE CURATED COLLECTION™

author + editor *extraordinaire*
ROBERTA NAAS

book design + *laborious layout*
KATRINA SOO HOO

book cover design + *digital artist*
ANDREA DIAZ

Geneva-born managing editor
NATHALIE GROLIMUND

editorial coordinator + *copyright police*
BIJAL PATEL

proofread *under the watchful eye of*
NICKY STRINGFELLOW
BIJAL PATEL

publisher assistant + *photocopy expert*
THOMAS FRIEBEL

production coordination
de.MO

book concept *conceived by*
PATRICE FARAMEH

published + produced by
farameh media llc
217 thompson street
10012 new york city usa
p +1 646 807 1810
f +1 646 417 7999
info@faramehmedia.com
www.faramehmedia.com

distributed worldwide by
daab media gmbh
scheidtweilerstr. 69
50933 cologne germany
p +49 221 690482 14
f +49 221 690482 29
mail@daab-media.com

We wish to thank all of the watch brands that participated in and submitted images and information for this luxurious ladies' book. Regretfully, not everyone could be included in this title, but we appreciate that there are a host of other wonderful, worthy brands for women that offer sport, daytime and simply classically elegant timepieces.

printed *on time in*
ITALY

ISBN 978-0-9830-8310-8

all rights reserved © 2011 FARAMEH MEDIA LLC *in* NEW YORK CITY *where time never stands still*
while we strive for utmost precision in every detail, we cannot be held responsible for any inaccuracies, or for any subsequent loss or damage arising.
any omissions for copy or credit are unintentional, and appropriate credit will be given in future editions if such copyright holders contact the publisher.
no part of this publication may be reproduced in any manner whatsoever without permission from the publisher. *period.*